Lecture Notes in Mathematics

Edited by A. Dold and B. Eckmann

709

Probability in Banach Spaces II

Proceedings of the Second International Conference on Probability in Banach Spaces, 18–24 June 1978, Oberwolfach, Germany

Edited by A. Beck

Springer-Verlag
Berlin Heidelberg New York 1979

Editor
Anatole Beck
Department of Mathematics
University of Wisconsin
Madison, WI 53706
USA

AMS Subject Classifications (1970): 28-XX, 28A40, 28A45, 46AXX, 46A05, 46B10, 60B10, 60B99, 60F05, 60F99

ISBN 3-540-09242-0 Springer-Verlag Berlin Heidelberg New York
ISBN 0-387-09242-0 Springer-Verlag New York Heidelberg Berlin

Printing and binding: Beltz Offsetdruck, Hemsbach/Bergstr.
2141/3140-543210

Introduction

The subject of Probability in Banach Spaces has now passed the quarter-century mark and is flourishing. It is over twenty years since the first paper linking geometric theory and probabilistic results, and nearly twenty since the first charaterization of a probabilistically signifficant category of spaces by convexity conditions. Since that time, the subject has continued to grow in this direction. Theorems shown to hold in uniformly convex spaces were strengthened to show that they held either exactly in B-convex spaces or exactly in super-reflexive spaces. Some of the Hilbert-space theorems were shown to hold exactly in type-2 spaces, etc. The vision of probability as being basically a subject grounded equally in geometry and in measure theory continues to be reinforced by the absence of any theorems which hold only in the real or complex numbers (except, trivially, for those which involve the multiplication of random variables). A few theorems have been found which hold exactly in finite-dimensional spaces, one for amarts and one for the strong law of large numbers, but these are not especially interesting in their own rights. Mostly, the theorems about real or complex numbers seem to flow from the fact that these are Hilbert spaces.

With this volume, Banach-space valued probability invades ever deeper into the territory of classical probability theory. The subject is now being accepted as being central to an understanding of many of the fundamental theorems there. It has truly come of age.

Anatole Beck,
Editor

CONTENTS

ON THE GENERAL CONVERSE CENTRAL LIMIT THEOREM IN BANACH SPACES

by

Alejandro de Acosta*
Instituto Venezolano de Investigaciones Científicas

1. Introduction. Let B be a separable Banach space, $\{X_{nj}\}$ a row-wise independent infinitesimal triangular array of B-valued random vectors, $S_n=\Sigma_j X_{nj}$. In a recent paper, de Acosta, Araujo and Giné [1] have obtained a general converse central limit theorem -necessary conditions for the weak convergence of $\{L(S_n)\}$-, valid for arbitrary B. In the present work we refine this result and obtain considerably more information on the limiting behavior of $\{L(S_n)\}$.

Our main result (Theorem 3.3) may be roughly stated as follows. Let $\{X_{nj}\}$ be an infinitesimal triangular array of B-valued random vectors and assume that $\{L(S_n)\}$ converges weakly. If $\{A_k:k=0,\ldots,m\}$ are appropriately chosen disjoint subsets of B, $X_{nj,k}$ is the truncation of X_{nj} at A_k and $S_{n;k}=\Sigma_j X_{nj,k}$, then $\{L(S_{n;0},\ldots,S_{n;m})\}$ converges weakly to a completely specified product measure on B^{m+1}. In less formal terms, the random vectors $\{S_{n;k}:k=0,\ldots,m\}$ are "asymptotically independent".

We remark that some of the basic results of the theory of B-valued stochastic processes with independent increments (see [2], Chapter IV, Theorem 5) follow easily from Theorems 3.3 and 4.1 of the present paper (together with some background results contained in [1]). We shall not carry out the details here.

---

* Research partially supported by CONICIT (Venezuela) Grant 51-26.S1.0893.

2. Preliminaries. We refer to [1] for the background of definitions and theorems which we will use. If X is a B-valued random vector (r. v.) and A is a Borel subset of B, we define X_A, the truncation of X at A, by $X_A = XI_{\{X \in A\}}$. We will consider truncations of a r.v. at disjoint sets $\{A_k : k=0,\ldots,m\}$; in this context we shall write X_k instead of X_{A_k}. If $\{X_{nj}\}$ is a triangular array, we will denote the truncation of X_{nj} at A_k by $X_{nj,k}$; the corresponding row sum will be written $S_{n;k} = \Sigma_j X_{nj,k}$.

3. Limiting behavior of the row sums of truncated random vectors. Our first result is proved by a classical method: a comparison argument showing that the difference between two sequences of characteristic functions tends to zero.

Proposition 3.1. Let $\{X_{nj}\}$ be an infinitesimal triangular array such that $\sup_n \Sigma_j P\{\|X_{nj}\| > \delta\} < \infty$ for some $\delta > 0$. Let $\{A_k : k=1,\ldots,m\}$ be disjoint Borel sets in B, $A_k \subset B_\delta^c$, and let $A_0 = (\cup_{k=1}^m A_k)^c$. Then for every $f_0,\ldots,f_m \in B'$,

$$\lim_n \left| \int \exp\{i\Sigma_{k=0}^m f_k(x_k)\} dL(S_{n;0},\ldots,S_{n;m})(x_0,\ldots,x_m) - \right.$$
$$\left. - \Pi_{k=0}^m \int \exp\{if_k(x)\} dL(S_{n;k})(x) \right| = 0.$$

Lemma 3.2. Let a_k be complex numbers, $|a_k| \le 1$ $(k=0,\ldots,m)$. Then

$$|\Sigma_{k=0}^m a_k - m - \Pi_{k=0}^m a_k| \le \Sigma_{\ell=0}^{m-1} \Sigma_{k=\ell+1}^m |a_k - 1||a_\ell - 1|.$$

Proof. $\Pi_{k=0}^m a_k + m - \Sigma_{k=0}^m a_k = \Sigma_{\ell=0}^{m-1}\{\Pi_{k=\ell}^m a_k - \Pi_{k=\ell+1}^m a_k - a_\ell + 1\}$

$$= \Sigma_{\ell=0}^{m-1}(\Pi_{k=\ell+1}^m a_k - 1)(a_\ell - 1).$$

Since $|\Pi_{k=\ell+1}^m a_k - 1| \le \Sigma_{k=\ell+1}^m |a_k - 1|$, we have

$$|\Pi_{k=0}^m a_k + m - \Sigma_{k=0}^m a_k| \le \Sigma_{\ell=0}^{m-1}\Sigma_{k=\ell+1}^m |a_k - 1||a_\ell - 1|. \quad \square$$

Proof of Proposition 3.1. Let L_n be the quantity the limit of which is taken in the statement of the proposition. Then

$$L_n = |E\exp\{i\Sigma_{k=0}^m f_k(S_{n;k})\} - \Pi_{k=0}^m E\exp\{if_k(S_{n;k})\}| =$$

$$= |E\exp\{i\Sigma_{k=0}^m \Sigma_j f_k(X_{nj,k})\} - \Pi_{k=0}^m E\exp\{i\Sigma_j f_k(X_{nj,k})\}|$$

$$=|\Pi_j E\exp\{i\Sigma_{k=0}^m f_k(X_{nj,k})\} - \Pi_j \Pi_{k=0}^m E\exp\{if_k(X_{nj,k})\}|$$

$$\leq \Sigma_j |E\exp\{i\Sigma_{k=0}^m f_k(X_{nj,k})\} - \Pi_{k=0}^m E\exp\{if_k(X_{nj,k})\}|.$$

Since A_k and $A_{k'}$ are disjoint for $k \neq k'$, it follows that for $k \neq k'$

$$f_k(X_{nj,k}) \cdot f_{k'}(X_{nj,k'}) = 0.$$

A simple calculation then yields

$$\exp\{i\Sigma_{k=0}^m f_k(X_{nj,k})\} = \Sigma_{k=0}^m \exp\{if_k(X_{nj,k})\} - m \quad \text{for all } n,j.$$

Using this fact and Lemma 3.2, we have

$$L_n \leq \Sigma_j \Sigma_{\ell=0}^{m-1} \Sigma_{k=\ell+1}^m |E\exp\{if_k(X_{nj,k})\}-1| \, |E\exp\{if_\ell(X_{nj,\ell})\}-1|$$

$$= \Sigma_{\ell=0}^{m-1} \Sigma_{k=\ell+1}^m \Sigma_j |E\exp\{if_k(X_{nj,k})\}-1| \, |E\exp\{if_\ell(X_{nj,\ell})\}-1|.$$

Now for $0 \leq \ell \leq m-1$, the infinitesimality of $\{X_{nj}\}$ implies that

$$\lim_n \sup_j |E\exp\{if_\ell(X_{nj,\ell})\}-1| = 0.$$

Also, for $1 \leq k \leq m$ we have

$$\Sigma_j |E\exp\{if_k(X_{nj,k})\}-1| \leq 2\Sigma_j P\{X_{nj} \in A_k\}$$

$$\leq 2\Sigma_j P\{\|X_{nj}\| > \delta\}.$$

It follows that for $0 \leq \ell \leq m-1$, $1 \leq k \leq m$,

$$\Sigma_j |E\exp\{if_k(X_{nj,k})\}-1| \, |E\exp\{if_\ell(X_{nj,\ell})\}-1| \leq$$

$$\leq \sup_j |E\exp\{if_\ell(X_{nj,\ell})\}-1| \Sigma_j |E\exp i\{f_k(X_{nj,k})\}-1|$$

$$\to 0 \quad \text{as } n \to \infty,$$

and therefore $\lim_n L_n = 0$. □

Remark. The particular case of Proposition 3.1 corresponding to $m=1$, $A_0 = B_\delta$, $A_1 = B_\delta^c$ may be used in §2 of [1] instead of Lemma 2.6 of [1]; this provides an alternative proof of the "asymptotic independence" of $S_{n,\delta}$ and $S_n^{(\delta)}$ (in the notation of [1]) without using the Lévy decomposition.

We recall some facts from [1] (Theorem 2.10). Let $\{X_{nj}\}$ be an infinitesimal triangular array and assume that $L(S_n - x_n)$ converges weakly to ν for some sequence $\{x_n\} \subset B$. Then there exist a Lévy measure μ and a centered Gaussian measure γ such that for every $\tau>0$, one has $\nu=\delta_{z_\tau}*\gamma*c_\tau\text{Pois}\mu$ for some $z_\tau\in B$; μ and γ are uniquely determined. Moreover, for every $\tau\in C(\mu)$ (the set of continuity radii of μ), $L(S_n-ES_{n,\tau})$ converges weakly to $\gamma*c_\tau\text{Pois}\mu$; here $S_{n,\tau}=\Sigma_j X_{nj}I_{\{X_{nj}\in B_\tau\}}$.

Theorem 3.3. Let $\{X_{nj}\}$ be an infinitesimal triangular array. Suppose $L(S_n-x_n)\to_w\nu$ for some sequence $\{x_n\}\subset B$, and let μ be the Lévy measure and γ the centered Gaussian measure associated to ν. Let $\tau\in C(\mu)$, $\{A_k:k=1,\ldots,m\}$ disjoint Borel sets such that $A_k\subset B_\tau^c$ and $\mu(\text{Bd}A_k)=0$ for $k=1,\ldots,m$, and let $A_0=(\cup_{k=1}^m A_k)^c$. Then

$$L(S_{n;0}-ES_{n,\tau},S_{n;1},\ldots,S_{n;m})\to_w(\gamma*c_\tau\text{Pois}(\mu|A_0))\otimes\text{Pois}(\mu|A_1)\otimes\ldots\otimes\text{Pois}(\mu|A_m).$$

Proof. By ([1], Theorem 2.10), $\Sigma_j L(X_{nj})|A_k\to_w\mu|A_k$ $(k=1,\ldots,m)$. By the Khinchine-Le Cam lemma ([1], Lemma 2.7), we have for the total variation norm $\|\cdot\|_v$

$$\begin{aligned}\|L(S_{n;k})-\text{Pois}(\Sigma_j L(X_{nj})|A_k)\|_v&=\|L(S_{n;k})-\text{Pois}(\Sigma_j L(X_{nj,k})\|_v\\&\le 2\Sigma_j(P\{X_{nj}\in A_k\})^2\\&\to 0 \text{ as } n\to\infty.\end{aligned}$$

Therefore $L(S_{n;k})\to_w\text{Pois}(\mu|A_k)$.

Let $T_n=S_n-S_{n;0}$. Then $L(T_n)\to_w\text{Pois}(\mu|A_0^c)$ (arguing as in the previous paragraph) and since $S_n-ES_{n,\tau}=(S_{n;0}-ES_{n,\tau})+T_n$, it follows that $\{L(S_{n;0}-ES_{n,\tau})\}$ is relatively compact and therefore so is $\{L(S_{n;0}-ES_{n,\tau},T_n)\}$. Now Theorem 2.2 of [1] and Proposition 3.1 imply that if λ is any subsequential limit of $\{L(S_{n;0}-ES_{n,\tau})\}$, then

$$\gamma*c_\tau\text{Pois}\mu=\lambda*\text{Pois}(\mu|A_0^c).$$

It follows that $\lambda=\gamma*c_\tau\text{Pois}(\mu|A_0)$ and therefore $L(S_{n;0}-ES_{n,\tau})\to_w\gamma*c_\tau\text{Pois}(\mu|A_0)$.

It follows from the preceding paragraphs that

$$L(S_{n;0}-ES_{n,\tau})\otimes L(S_{n;1})\otimes\ldots\otimes L(S_{n;m})\to_w$$

$$\to_w (\gamma * c_\tau \mathrm{Pois}(\mu|A_0)) \otimes \mathrm{Pois}(\mu|A_1) \otimes \ldots \otimes \mathrm{Pois}(\mu|A_m).$$

On the other hand, $\{L(S_{n;0}-ES_{n,\tau}, S_{n;1}, \ldots, S_{n;m})\}$ is relatively compact because each sequence of marginal probabilities is relatively compact; the conclusion follows now from Proposition 3.1. □

Corollary 3.4. Let $\nu=\delta_z * \gamma * c_\tau \mathrm{Pois}\mu$ (with $\mu(\{0\})=0$) be an infinitely divisible measure on B.

(1) If $\{X_{nj}\}$ is an infinitesimal triangular array such that $L(S_n - x_n) \to_w \nu$ for some sequence $\{x_n\} \subset B$, A is an open subset of B and $P\{X_{nj} \in A\}=0$ for all n,j, then $\mu(A)=0$.

(2) Conversely, if C is a closed subset of B such that $C \subset B_\delta^c$ for some $\delta>0$ and $\mu(C)=0$, then there exists an infinitesimal triangular array $\{X_{nj}\}$ such that $P\{X_{nj} \in C\}=0$ for all n,j and $L(S_n) \to_w \nu$.

Proof. (1): Follows at once from [1], Theorem 2.10.

(2) Let $\{Y_{nj}\}$ be an infinitesimal triangular array such that $L(T_n) \to_w \nu$, where $T_n = \Sigma_j X_{nj}$; say, take $\{Y_{nj}: j=1,\ldots,n\}$ independent and $L(Y_{nj}) = \nu_n$, where ν_n is the n-th convolution root of ν.

Since $\mu(\mathrm{Bd}C)=0$ and $\mu=\mu|C^c$, it follows from Theorem 3.3 that if we define $X_{nj} = Y_{nj} I_{\{Y_{nj} \in C^c\}}$, then $\{X_{nj}\}$ satisfies the desired condition. □

Corollary 3.5. Let ν be as in Corollary 3.4. Then μ has bounded support if and only if there exists an infinitesimal uniformly bounded triangular array $\{X_{nj}\}$ such that $L(S_n) \to_w \nu$.

4. The limiting distribution of the number of visits. For a>0, let us denote by $\pi(a)$ the Poisson distribution with parameter a on R^1 ($\pi(a) = \delta_0$ if a=0). Let A be a Borel set in B satisfying $d(0,A)>0$ and let $\phi^{(n)}$ be the number of visits to A by the random vectors in the n-th row of the triangular array $\{X_{nj}\}$; that is, $\phi^{(n)} = \Sigma_j I_A(X_{nj})$. If the assumptions of the general converse central limit theorem ([1], Theorem 2.10) are

satisfied and $\mu(\mathrm{Bd}A)=0$, then $L(\phi^{(n)})$ converges weakly to $\pi(\mu(A))$; here μ is the Lévy measure associated to $\{X_{nj}\}$. In fact, more is true; if $\|\cdot\|_v$ denotes the total variation norm, then we have

<u>Theorem 4.1.</u> Let $\{X_{nj}\}$ be an infinitesimal triangular array. Suppose $L(S_n-x_n)\to_w \nu$ for some sequence $\{x_n\}\subset B$ and let μ be the Lévy measure associated to ν. Let $\delta>0$, $\{A_k:k=1,\dots m\}$ be disjoint Borel sets such that $A_k\subset B_\delta^c$ and $\mu(\mathrm{Bd}A_k)=0$ for $k=1,\dots,m$. Let $\phi_k^{(n)}=\Sigma_j I_{A_k}(X_{nj})$. Then

$$\lim_n\| L(\phi_1^{(n)},\dots,\phi_m^{(n)})-\otimes_{k=1}^m \pi(\mu(A_k))\|_v=0.$$

<u>Proof.</u> Let $V_{nj}=(I_{A_1}(X_{nj}),\dots,I_{A_m}(X_{nj}))$; then $\{V_{nj}\neq 0\}\subset\{\|X_{nj}\|>\delta\}$. By the Khinchine- Le Cam lemma ([1], Lemma 2.7),

$$\begin{aligned}\| L(\phi_1^{(n)},\dots,\phi_m^{(n)})-\mathrm{Pois}(\Sigma_j L(V_{nj}))\|_v&=\| L(\Sigma_j V_{nj})-\mathrm{Pois}(\Sigma_j L(V_{nj}))\|_v\le 2\Sigma_j (P\{V_{nj}\neq 0\})^2\\ &\le \sup_j P\{\|X_{nj}\|>\delta\}\Sigma_j P\{\|X_{nj}\|>\delta\}\\ &\to 0 \text{ as } n\to\infty.\end{aligned}$$

Now $L(V_{nj})=(1-\Sigma_{k=1}^m P\{X_{nj}\in A_k\})\delta_0+\Sigma_{k=1}^m P\{X_{nj}\in A_k\}\delta_{e_k}$ (here $0\in R^m$ and $e_1,\dots,e_m$ are the unit vectors in R^m) and so

$$\Sigma_j L(V_{nj})=(\Sigma_j(1-\Sigma_{k=1}^m P\{X_{nj}\in A_k\}))\delta_0+\Sigma_{k=1}^m(\Sigma_j P\{X_{nj}\in A_k\})\delta_{e_k}$$

and $\mathrm{Pois}(\Sigma_j L(V_{nj}))=\mathrm{Pois}(\Sigma_{k=1}^m(\Sigma_j P\{X_{nj}\in A_k\})\delta_{e_k})$.

By Theorem 2.10 of [1],

$$\begin{aligned}\|\Sigma_{k=1}^m(\Sigma_j P\{X_{nj}\in A_k\})\delta_{e_k}-\Sigma_{k=1}^m \mu(A_k)\delta_{e_k}\|_v &\le\\ \le \Sigma_{k=1}^m|\Sigma_j P\{X_{nj}\in A_k\}-\mu(A_k)|&\\ \to 0 \text{ as } n\to\infty.&\end{aligned}$$

It follows that $\|\mathrm{Pois}(\Sigma_j L(V_{nj}))-\mathrm{Pois}(\Sigma_{k=1}^m \mu(A_k)\delta_{e_k})\|_v\to 0$ as $n\to\infty$. An elementary calculation shows that

$$\mathrm{Pois}(\Sigma_{k=1}^m a_k\delta_{e_k})=\otimes_{k=1}^m \pi(a_k)$$

for $a_k\ge 0$ $(k=1,\dots,m)$, completing the proof. □

REFERENCES

[1] de Acosta, A., Araujo, A., and Giné, E. (1977). On Poisson measures, Gaussian measures and the central limit theorem in Banach spaces. To appear in Advances in Probability, Vol. 4.

[2] Gihman, I.I., and Skorohod, A.V. (1975). The Theory of Stochastic Processes, Vol. II. Springer-Verlag, Berlin and New York.

Stable Processes with Continuous Sample Paths

A. Araujo and M. B. Marcus[(1)]

1. Introduction

Stable processes have received a great deal of attention in classical probability in the study of real valued stochastic processes with independent increments and it is well known that in this case, only a stable process of exponent 2, a time changed brownian motion, has continuous sample paths. Nevertheless there are many non-trivial stable processes of exponent $p < 2$ with continuous sample paths. Let $\{a_k\} \in \ell^1$ and let $\{\theta_k\}$ be independent copies of θ where θ is given by its characteristic function

$$E[e^{iu\theta}] = e^{-|u|^p}, \quad -\infty < u < \infty.$$

(Throughout this paper $\{\theta_k\}$ and θ will be defined in this way.) Then for $1 < p \leq 2$ the random Fourier series $\Sigma\, a_k \theta_k e^{ikt}$, $t \in [-1/2, 1/2]$ converges uniformly a.s. and is a stable process. (See also [3]). Here we use the customary definition of a stable process of exponent p. Let X be a random variable with values in $C[-1/2,1/2]$; then X is a stable process of exponent p

(1) This research was supported, in part, by a grant from the National Science Foundation, U.S.A.

whenever for X_1 and X_2 independent copies of X and any $\alpha,\beta > 0$ there exists a $r > 0$ such that

$$\mathcal{L}(\alpha X_1 + \beta X_2) = \mathcal{L}(rX).$$

We will give two sufficient conditions for the continuity of the sample paths of stable processes. One for a class of stationary stable processes of the form

$$\int_{-\infty}^{\infty} e^{it\lambda} M(d\lambda), \quad t\in[-1/2,1/2] \tag{1.1}$$

and the other for a more general class of processes represented by

$$\int_B xM(dx); \tag{1.2}$$

where M is an independently scattered random stable measure of exponent p on the real line in (1.1) and on the unit ball B of $C[-1/2,1/2]$ in (1.2). The definition and proof of existence of these random integrals is given in Section 2. The sufficient condition for continuity in the stationary case is given in Section 3. It is essentially an adaptation to the stable case of a recent result of Fernique [4] on the sample path continuity of a certain class of weakly stationary second order processes. In Section 4 we give a sufficient condition for the sample path continuity of more general stable processes. We present this result in two ways. The first (Theorem 4.1), which applies to certain processes of the type (1.2), is an immediate consequence of extending the definition of random integrals on spaces of stable type given in [8] following an idea introduced by Zinn [12]. The second approach (Theorem 4.3) is only slightly different. It establishes the existence of Levy measures of stable measures on the space of continuous processes. It is a corollary of the general development of [1] and E. Giné has informed us that he also has a proof of our Theorem 4.3. However, even though these theorems follow readily from established theory we show by examples that they are close to the best possible results in certain cases.

The results of Section 3 show that there are interesting differences between stationary Gaussian processes (the case $p = 2$) and other stationary stable processes. These will be discussed at the end of Section 3.

We recall some standard notation. Let (S,Σ) be a measurable space, $s\in S$ and m a measure on (S,Σ). Let F be a separable Banach space with norm $\| \ \|_F$ and f a function from S into F. By $L^p(S,\Sigma,m;\ F)$ we mean those functions f for which

$$\int_S \|f(s)\|^p m(ds) < \infty.$$

Let $(\Omega,\mathcal{F},P)$ denote a probability space. Let $X\colon \Omega \to F$, i.e. X is an F valued random variable. By $L^p(\Omega,\mathcal{F},P;\ F)$ we mean those F valued random variables X for which

$$E\|X\|_F^p < \infty.$$

2. Random Integrals

In this section we shall explain what we mean by random integrals of the type (1.1) and (1.2). To begin we pursue an idea of Zinn [12] (see also [6]) and extend the notion of spaces of stable type p to operators of stable type p. Let E and F be separable Banach spaces and let $T: E \to F$ be a continuous linear operator.

Definition 2.1. T is said to be of stable type p, $0 < p \leq 2$, if for each q, $0 < q < p$ if $p < 2$ and $0 < q \leq 2$ for $p = 2$, there exists a constant C such that for all integers n and any $x_1,\ldots,x_n \in E$

$$(2.1) \qquad (E\|\sum_{i=1}^{n} \theta_i(Tx_i)\|_F^q)^{1/q} \leq C(\sum_{i=1}^{n} \|x_i\|_E^p)^{1/p},$$

where $\| \ \|_E$ and $\| \ \|_F$ indicate the Banach space norms on E and F respectively. In the case that $E = F$ and T is the identity operator we say that the Banach space E is of stable type p.

By familiar arguments Definition 2.1 is equivalent to the following: T is of stable type p, $0 < p \leq 2$ if for all sequences $\{x_i\}$ of elements of E for which $\Sigma\|x_i\|_E^p < \infty$ the series $\Sigma\, \theta_i(Tx_i)$ converges a.s. in F.

We now follow [8] (see also [11]). Let (S,Σ) be a measurable space. A mapping M from Σ into real valued random variables on a probability space $(\Omega,\mathcal{F},P)$ is said to be an independently scattered random measure if for any pair-wise disjoint $A_1,A_2,\ldots \in \Sigma$ the random variables $M(A_1),M(A_2),\ldots$ are independent and

$M(\cup A_i) = \Sigma M(A_i)$ where the series on the right converges in probability.

Let m be a non-negative finite measure on (S,Σ). Fix p, $0 < p \leq 2$ for each $A \in \Sigma$ we define the real valued random variable $M(A)$ by

$$E \exp(iuM(A)) = \exp(-m(A)|u|^p), \quad -\infty < u < \infty.$$

(Note that $\mathcal{L}(M(A)) = \mathcal{L}(m^{1/p}(A)\theta)$. By the Kolmogorov consistency theorem we see that M is an independently scattered random stable measure of exponent p on (S,Σ). We call m the control measure of M.

If $f\colon S \to E$ is a simple function, i.e. $f = \sum_{i=1}^{n} x_i I_{A_i}$, where $A_i \in \Sigma$ are pairwise disjoint and $x_i \in A_i$ (I_A denotes the indicator function of the set A) we set

$$(2.2) \qquad \int_S Tf(s)M(ds) = \sum_{i=1}^{n} Tx_i M(A_i).$$

In this $\int TfdM$ is a stable F-valued random vector. For f simple and in $L^p(S,\Sigma,m;\, E)$ and for $q < p$ the map $f \to \int Tf(s)M(ds)$ is a linear operator with values in $L^q(\Omega,\mathcal{F},P;\, F)$. By (2.1) we have

$$(2.3) \qquad (E\|\int Tf(s)M(ds)\|_F^q)^{1/q} = (E\|\Sigma\, Tx_i M(A_i)\|_F^q)^{1/q}$$

$$\leq C(\Sigma\|x_i\|_E^p m(A_i))^{1/p}$$

$$= C(\int\|f(s)\|^p m(ds))^{1/p}.$$

Since the simple functions are dense in $L^p(S,\Sigma,m; E)$ (there exists a unique extension of this operator onto the whole of $L^p(S,\Sigma,m; E)$. This extension will also be denoted by $\int TfdM$ and also satisfies (2.3). Thus we have proved the following theorem.

Theorem 2.2: Let $T: E \to F$ be of stable type p, $0 < p \leq 2$ and let M be an independently scattered random stable measure of exponent p on (S,Σ) with control measure m. Then for each q, $0 < q < p < 2$ if $p < 2$ and $0 < q \leq 2$ for $p = 2$, there exists a linear map

$$(2.4) \qquad L^p(S,\Sigma,m; E) \ni f \to \int_S TfdM \in L^q(\Omega, \mathcal{F}, P; F)$$

the values of which are stable random vectors in F satisfying

$$(2.5) \qquad (E\|\int TfdM\|_F^q)^{1/q} \leq C(\int \|f\|_E^p)\,dm)^{1/p}$$

for some finite constant C independent of f.

For later use we mention the following well known lemma (see Lemma 4.4 [9]).

Lemma 2.3. Let $\{a_i\}$ be a sequence of real numbers and $\{\theta_i\}$ independent copies of θ, a symmetric stable random variable of index p, $1 < p \leq 2$, then

$$E(\Sigma\, a_i^2 \theta_i^2)^{1/2} \leq C(\Sigma |a_i|^p)^{1/p}$$

where C is a constant depending only on p.

3. Stationary Stable Process

Consider the random integral

$$X(t) = \int_{-\infty}^{\infty} e^{it\lambda} M(d\lambda), \quad t \in [-1/2,1/2] \tag{3.1}$$

where M is an independently scattered random stable measure of exponent p on (R,Σ), the real line with the Borel field. In the case $p = 2$ (3.1) is a special case of a general representation for weakly stationary second order processes. In the case $p < 2$ the existence of this integral has also been shown (see for example [2],[3]). We will define it here by means of Theorem 2.2. Let $S = R$ and $E = F = L^2([-1/2,1/2],\mu)$ where μ is Lebesgue measure. We choose for m any finite positive measure and let M be the corresponding independently scattered random stable measure of exponent p with control measure m. For $\lambda \in R$ the function $f: \lambda \to e^{it\lambda}$ is in $L^2([0,1]; \mu)$ and $f \in L^p(R,\Sigma,m; E)$. Then, since L^2 is of stable type p, $p \leq 2$, the random integral (3.1) exists and furthermore

$$E\|X(t)\|_2^q < \infty; \tag{3.2}$$

(we denote $\| \ \|_E$ by $\| \ \|_2$ in this case). We choose a separable version for (3.1) and denote it by $X = \{X(t), t \in [-1/2,1/2]\}$ and note that $X(t)$ is a stable random variable with

$$E[e^{iuX(t)}] = e^{-|u|^p m(R)}$$

and

$$E[e^{iu(X(t+h)-X(t))}] = e^{-|u|^p \int_{-\infty}^{\infty} |\sin \frac{\lambda h}{2}|^p m(d\lambda)} \quad . \tag{3.3}$$

By considering the characteristic function of the finite joint distributions of X we see that X is a stationary process and (3.2) implies

$$(3.4) \qquad E|X(t)|^q < \infty.$$

(Recall that we take $0 < q < p$ for $1 < p < 2$ and $0 < q \leq 2$ for $p = 2$). Let

$$(3.5) \qquad \sigma_p(h) = \left(\int_{-\infty}^{\infty} \left|\sin \frac{\lambda h}{2}\right|^p m(d\lambda)\right)^{1/p}.$$

By (3.3) we have that $\mathcal{L}(X(t+h)-X(t)) = \mathcal{L}(\sigma_p(h)\theta)$, so

$$(3.6) \qquad (E|X(t+h)-X(t)|^q)^{1/q} = C_{p,q}\sigma_p(h)$$

where $C_{p,q} = (E|\theta|^q)^{1/q}$. By taking $q \geq 1$ we have that $\sigma_p(h)$ is a metric on the set of random variables $X = \{X(t), t\in[-1/2,1/2]\}$. In keeping with the terminology in the case $p = 2$ we will also refer to m as the spectral measure of X.

Let $\mu_{\sigma_p}(\epsilon) = \lambda\{h\in[-1,1], \sigma_p(h) < \epsilon\}$ where λ is Lebesgue measure. Define

$$\overline{\sigma_p(h)} = \sup\{y | \mu_{\sigma_p}(y) < h\}$$

and set $\hat{\sigma}_p = \sup_{h\in[-1,1]} \sigma(h)$. We see that $\overline{\sigma}_p$ is a non-decreasing function on $[0,2]$ and $0 \leq \overline{\sigma}_p \leq \hat{\sigma}_p$. The function $\overline{\sigma}_p$ is called the non-decreasing rearrangement of σ_p. We define the integral

$$I(\sigma_p) = I(\sigma_p(u)) = \int_0^2 \frac{\overline{\sigma_p(u)}}{u(\log \frac{16}{u})^{1/2}} du.$$

The following theorem compliments Fernique's theorem in [4].

Theorem 3.1: Let X be a separable stationary stable process with exponent $1 < p \leq 2$ and with spectral measure m as given in (3.1). Let $\{a_k\}$ be a sequence of positive real numbers increasing to infinity. If $I(\sigma_p) < \infty$ the processes

$$X_k(t) = \int_{-a_k}^{a_k} e^{it\lambda} M(d\lambda), \quad t \in [-1/2, 1/2]$$

have continuous sample paths a.s. and converge uniformly to X a.s. Therefore X also has continuous sample paths a.s.

Note that for $p = 2$, $X(t)$ is a real valued stationary Gaussian process on $[-1/2,1/2]$ and all such processes have a representation as in (3.1). In this case the condition $I(\sigma_2) < \infty$ is the well known Dudley-Fernique necessary and sufficient condition for the continuity of a stationary Gaussian process.

Proof: Assume that $m([-1,1]) \neq 0$. For each integer $k \geq k_0$ let $\{A_{k,i}\}$ be a partition of R into disjoint intervals such that for each i there exists a $\lambda_{k,i} \in A_{k,i}$ and $|\lambda - \lambda_{k,i}| \leq 1/k$ for all $\lambda \in A_{k,i}$. Consider

$$(3.7) \qquad Y_k(t) = \sum_i e^{it\lambda_{k,i}} M(A_{k,i})$$

and let $4\delta = \int_{-1}^{1} |\lambda|^p m(d\lambda)$. We will obtain two inequalities:

$$(3.8)\qquad (E|Y_k(h)-Y_k(0)|^q)^{1/q} \le C_1\sigma_p(h)$$

for $k > [1/\delta] + 1 = k_0$, and

$$(3.9)\qquad (E|X(t)-Y_k(t)|^q)^{1/q} \le \frac{C_2 t}{k}$$

where C_1 and C_2 are constants independent of k.

To obtain (3.8) we note that

$$(3.10)\qquad (E|Y_k(h)-Y_k(0)|^q)^{1/q} \le (E|X(0)-X(h)|^q)^{1/q}$$
$$+ (E|X(h)-Y_k(h)|^q)^{1/q} + (E|X(0)-Y_k(0)|^q)^{1/q} .$$

The last term in (3.10) is zero and

$$(3.11)\qquad (E|X(h)-Y_k(h)|^q)^{1/q} = C_{p,q}(\sum_i \int_{A_{k,i}} |\sin \frac{h(\lambda-\lambda_{k,i})}{2}|^p m(d\lambda))^{1/p}$$
$$\le \frac{C_{p,q} m^{1/p}(R)h}{2k}$$

Therefore by (3.6) and (3.10) we have

$$(3.12)\qquad (E|Y_k(h)-Y_k(0)|^q)^{1/q} \le C[\sigma_p(h) + \frac{h}{k}]$$

for some constant C. Since

$$(3.13)\qquad \sigma_p(h) \ge (\int_{-1}^{1} |\sin \frac{\lambda h}{2}|^p m(d\lambda))^{1/p} \ge \delta h$$

we can use (3.13) in (3.12) to obtain (3.8). Inequality (3.9) already appears in (3.11).

The next step is to find a uniform bound for $E[\sup_{t\in[-1/2,1/2]} |X_k(t)|]$. We follow the approach of [10]. Let

$\{\epsilon_{k,i}\}$ be a Rademacher sequence independent of $\{M(A_{k,i})\}$. Clearly (3.7) is stochastically equivalent to

$$\sum_i e^{it\lambda_{k,i}} \epsilon_{k,i} M(A_{k,i}). \tag{3.14}$$

Let $(\Omega_1, \mathcal{F}_1, P_1)$ denote the probability space of $\{M(A_{k,i})\}$ and $(\Omega_2, \mathcal{F}_2, P_2)$ the probability space of $\{\epsilon_{k,i}\}$ and denote the corresponding expectation operators by E_1 and E_2. The series (3.14) is defined on the probability space $(\Omega_1 \times \Omega_2, \mathcal{F}_1 \times \mathcal{F}_2, P_1 \times P_2)$. We shall refer to this space as $(\Omega, \mathcal{F}, P)$ and denote the corresponding expectation operator by E. Consider

$$Y_k(t,\omega_1) = \sum_i e^{it\lambda_{k,i}} M(A_{k,i},\omega_1), \quad t \in [-1/2, 1/2], \quad \omega_1 \in \Omega_1.$$

By (22) of [10] we have

$$E_2\Big[\sup_{t\in[-1/2,1/2]} |Y_k(t,\omega_1)|\Big] \le D\Big[\big(\sum_i |M(A_{k,i},\omega_1)|^2\big)^{1/2} + \tilde{I}(\tau_p(u,\omega_1))\Big] \tag{3.15}$$

where D is a constant,

$$\tau_p(u,\omega_1) = \Big(\sum_i |M(A_{k,i},\omega_1)|^2 \sin^2 \frac{\lambda_{k,i} u}{2}\Big)^{1/2}$$

and

$$\tilde{I}(\tau_p(u,\omega_1)) = \tilde{I}(\tau_p)$$

$$= \int_0^2 \frac{\overline{\int_0^s \tau_p(u,\omega_1)\,du}}{s^2 \left(\log \frac{16}{s}\right)^{1/2}} ds.$$

By Lemma 2.3 we have

$$E\left(\sum_i |a_i\theta_i|^2\right)^{1/2} \leq C\left(\sum_i |a_i|^p\right)^{1/p} \tag{3.16}$$

and by Lemma 4.3 [9]

$$E_1\left(\int_0^s \overline{\tau_p(u,\omega_1)}\,du\right) \leq \int_0^s \overline{E_1\tau(u,\omega_1)}\,du. \tag{3.17}$$

We apply E_1 to each side of (3.15) and using (3.6), (3.16) and (3.17) we get

$$E[\sup_{t\in[-1/2,1/2]} |Y_k(t)|] \leq D'[m^{1/p}(R)$$

$$+ \tilde{I}((E|Y_k(h)-Y_k(0)|^q)^{1/q}]$$

for some constant D' independent of k. We now use (3.8) and inequality (16) of [10] to obtain

$$E[\sup_{t\in[-1/2,1/2]} |Y_k(t)|] \leq 2D'[m^{1/p}(R) + I(\sigma_p)]$$

and this holds for all k. Following Fernique's argument prior to his Proposition 2.3 [4] we see that the separability of X together with (3.9) gives

$$E[\sup_{t\in[-1/2,1/2]} |X(t)|] \leq 2D'[m^{1/p}(R) + I(\sigma_p)]. \tag{3.18}$$

(That (3.18) is equivalent to Proposition 2.3 [4] follows from Lemma 5 [10].)

Now consider $X_k(t)$. We have

$$E|X_k(t+h)-X_k(t)|^q \leq C_k|h|^q$$

for some constant C_k. Therefore by Kolmogorov's theorem ([5], pg. 171), $X_k(t)$ has continuous sample paths a.s. Also we note that the increments $\{X_{j+1}(t)-X_j(t)\}$ are independent. Using (3.18) and Lévy's inequality we get

$$E\{\sup_{j\leq k}\ \sup_{t\in[-1/2,1/2]} |X(t)-X_j(t)|\} \tag{3.19}$$

$$\leq 4D'\{m^{1/p}([-a_k,a_k]^c) + I((\int_{\lambda>|a_k|} |\sin \tfrac{\lambda h}{2}|^p m(d\lambda))^{1/p})\}.$$

This inequality enables us to complete the proof of the theorem under the assumption $m([-1,1]) \neq 0$.

We now remove this restriction. Without loss of generality we can assume that m is not concentrated at zero. Let h_0 be such that

$$(\int_{-1/h_0}^{1/h_0} |\lambda|^p m(d\lambda))^{1/p} = 4\delta' > 0.$$

Then

$$\sigma_p(h) \geq (\int_{-1/h_0}^{1/h_0} |\sin \tfrac{\lambda h}{2}|^p m(d\lambda))^{1/p} \geq \delta' h$$

for $0 \leq h \leq h_0$. Now consider $X(t)$ and $X_k(t)$ as above but with $t\in[-\frac{h_0}{2},\frac{h_0}{2}]$. The same analysis as above show that all conclusions hold for this process as long as

$$I(\sigma_p(h_0,u)) = \int_0^{2h_0} \frac{\overline{\sigma_p(h_0,u)}}{u(\log \frac{16h_0}{u})^{1/2}}du < \infty$$

where $\overline{\sigma_p(h_0,u)}$ is the non-decreasing rearrangement of $\sigma_p(h)$ with respect to $[-h_0,h_0]$. (In this notation $I(\sigma_p)$ defined above is $I(\sigma_p(1,u))$). By Lemma 4 and 5 and (13) and (14) of [10] we see that if $I(\sigma_p(h_0,u)) < \infty$ for some $h_0 > 0$ then it is finite for all $h_0 > 0$. Thus $I(\sigma_p) < \infty$ implies $I(\sigma_p(h_0,u)) < \infty$ which proves the theorem for $\{X(t), t\in[-\frac{h_0}{2},\frac{h_0}{2}]\}$. Then, by stationarity the theorem is also true for $t\in[-1/2,1/2]$. This completes the proof of the theorem.

There are some interesting differences between results on the sample path continuity of stationary Gaussian processes (the case $p = 2$) and other stable processes. Recall that σ_p defined in (3.5) is a metric on $X = \{X(t), t\in[-1/2,1/2]\}$. Let $N_{\sigma_p}(X,\varepsilon)$ denote the minimum number of balls of radius ε in the σ_p metric necessary to cover X. By Lemma 3 [10] we have

$$(3.20) \qquad I(\sigma_p) < \infty \iff \int_0 (\log N_{\sigma_p}(X,\varepsilon))^{1/2} d\varepsilon < \infty.$$

Therefore, Theorem 3.1 extends Dudley's sufficient condition for the continuity of Gaussian processes to stable non-Gaussian processes of the type given in (3.1). However, the right side of (3.20) is also a sufficient condition for the sample path continuity of all Gaussian processes and although it makes sense to consider it as possibly a sufficient condition for the continuity of all stable processes of exponent p, it is not. To see this recall that for Y a stable process of exponent p with independent stationary increments we have

$$(E|Y(t+h)-Y(t)|^q)^{1/q} = Ch^{1/p}$$

$(1 \leq q < p)$ for some constant C and $I(\sigma_p) < \infty$ in this case. (The equivalence (3.20) continues to hold for processes with stationary increments.)

A second difference is that whereas the integrals in (3.20) are necessary and sufficient for the continuity of stationary Gaussian processes, Gilles Pisier has informed us that he has a counterexample to the conjecture that they are also necessary and sufficient for the continuity of processes of the type (3.1) for $1 < p < 2$.

Finally, another problem is that whereas all stationary Gaussian processes can be represented by a random integral of the type given in (3.1) we do not know if the same is true for stationary stable processes of exponent $p < 2$.

If m is discrete and places mass $|a_k|^p$ at $\lambda = k$ (3.1) is a random Fourier series

(3.21) $$\sum_{k=-\infty}^{\infty} a_k \theta_k e^{ikt}, \quad t \in [-1/2, 1/2]$$

(for m discrete Theorem 3.1 is given in [9].) It follows from Theorem 3.1 and the method of Chapter 4 §3 [7] that a sufficient condition for the uniform convergence a.s. of (3.21) is

(3.22) $$\sum_k \frac{(\sum_{j=k}^{\infty} |a_j|^p)^{1/p}}{k(\log k)^{1/2}} < \infty .$$

We shall refer to this in Section 4.

4. General Case

Let (S,d) be a compact metric space and let F denote the Banach space $\{C(S),\| \|_\infty\}$, the continuous functions on S with the usual sup norm. Let τ be a continuous metric or pseudo-metric on S. We define

$$\mathrm{Lip}(\tau) = \{x\in C(s) \mid \sup_{\substack{s,t\in S\\ s\neq t}} \frac{x(s)-x(t)}{\tau(x,t)} < \infty\}$$

and for fixed $t_0 \in S$ the norm

$$(4.0) \qquad |||x|||_\tau = |x(t_0)| + \sup_{\substack{s,t\in S\\ s\neq t}} \left|\frac{x(s)-x(s)}{\tau(s,t)}\right|.$$

As above we denote by $N_\tau(S,\varepsilon)$ the minimum number of balls of radius ε in the τ pseudo-metric that is necessary to cover S. We will be concerned with these τ for which

$$(4.1) \qquad \int_0 (\log N_\tau(S,\varepsilon))^{1/2} d\varepsilon < \infty.$$

Let m be a finite positive measure on the unit ball B of $C(S)$. We prove the following theorem.

Theorem 4.1. For $0 < p \leq 2$ assume that

$$\int_B |||x|||_\tau^p m(dx) < \infty$$

and that τ satisfies (4.1). Let M be an independently scattered random stable measure of exponent p with control measure m. Then

$$\int_B x\, M(dx) \in L^q(\Omega, \quad , P; F)$$

where $q < p$ for $0 < p < 2$ and $q \leq 2$ for $p = 2$. (i.e. $\int_B x\, dM$ is a stable stochastic process of exponent p with continuous sample paths and $E\|\int_B x\, dM\|_\infty^q < \infty$).

<u>Proof</u>: Following [12] (see also [6]) we know that given τ satisfying (4.1) we can find a τ' such that

$$\lim_{\substack{s \to t \\ s \neq t}} \frac{\tau(s,t)}{\tau'(s,t)} = 0$$

and

$$(4.2) \qquad \int_0 (\log N_{\tau'}(S,\varepsilon))^{1/2} d\varepsilon < \infty.$$

We now consider the separable Banach space $E = \{Lip(\tau), \||\ \||_{\tau'}\}$ and show that the identity operator I from E into F is of stable type p. Let $\{x_i\}$ be a sequence of elements of E with $\Sigma \||x_i\||_{\tau'}^p < \infty$; we will show that this implies $\Sigma\, \theta_i x_i$ converges a.s. in F. The argument is standard. Let $(\Omega_1, \mathcal{F}_1, P_1)$ and $(\Omega_2, \mathcal{F}_2, P_2)$ be the probability spaces introduced in the proof of Theorem 3.1. For fixed $\omega_1 \in \Omega_1$ we consider the subgaussian process

$$Z(t,\omega_1) = \sum_i \varepsilon_i |\theta_i(\omega_1)| x_i(t).$$

We have

$$\{E_2|Z(t,\omega_1)-Z(s,\omega_1)|^2\}^{1/2}$$

$$\leq (\sum_i|\theta_i(\omega_1)|^2|x_i(t)-x_i(s)|^2)^{1/2}$$

$$\leq \tau'(s,t)(\sum_i|\theta_i(\omega_1)|^2\| x_i\|^2_{\tau'})^{1/2}$$

Taking (4.2) into account we know that $Z(t,\omega_1)$ converges uniformly a.s. (P_2) for those values of ω_1 for which $\Sigma|\theta_i(\omega_1)|^2\| x_i\|^2_{\tau'} < \infty$. By the three series theorem this series converges a.s. (P_1). Therefore $Z(t,\omega_1)$ converges a.s. (P_2) for $\omega_1\epsilon\bar{\Omega}_1 \subset \Omega_1$ for which $P(\bar{\Omega}_1) = 1$. It follows by Fubini's theorem that $\Sigma\ \theta_i x_i$ converges a.s. in C(S). Thus I is of stable type p.

The proof now follows from Theorem 2.2 in which we take (S,Σ) to be F with the Borel field generated by the norm.

We present as Theorem 4.3 an alternate proof of Theorem 4.1. Let R be a positive symmetric Borel measure on a separable Banach space $(G,\|\ \|_G)$ such that $R(\{0\}) = 0$ and $R(x\epsilon G,\|\ \|_G > a) < \infty$ for all $a > 0$. R is called a Lévy measure if the function φ defined on $f\epsilon G^*$ by

(4.3) $$\varphi(f) = \exp \int_G [\exp(if(x)-1)R(dx)$$

is the characteristic functional of a probability measure P_R on G. If R has the form

(4.4) $$R(dx) = \frac{C}{r^{1+p}}drU(du)$$

for some constant $C > 0$ and $0 < p < 2$ where $r = \|x\|_G$ and $u = x/\|x\|_G$ for $x \neq 0$ then the corresponding measure P_R on G, if it exists, is a stable measure of exponent p. (Note that U is a measure on the unit ball of G.) We will use the following result from [1].

<u>Theorem 4.2</u>. Let $(B_1, \| \ \|_1)$ and $(B_2, \| \ \|_2)$ be two separable Banach spaces and $T: B_1 \to B_2$ be a continuous linear operator of stable type 2. Then

(4.5) $$\int_{B_1} \min(1, \|x\|_1^2) R(dx) < \infty$$

is a sufficient condition for $R \circ T^{-1}$ to be a Lévy measure on B_2.

Recall that we designate by E and F the separable Banach spaces $\{\mathrm{Lip}(\tau), ||| \ |||_{\tau}, \}$ and $\{C(S), \| \ \|_{\infty}\}$. We can now prove

<u>Theorem 4.3</u>: Let R be a positive symmetric Borel measure on F of the type given in (4.4). Suppose that (4.1) holds and that

(4.6) $$\int ||| u |||_{\tau}^{p} U(du) < \infty.$$

Then R is the Lévy measure of a stable measure of exponent p on F. (Here $||| \ |||_{\tau}$ is the function on F given in (4.0).).

<u>Proof</u>: The proof follows from Theorem 4.2 along similar lines as the proof of Theorem 4.1. We have already shown that the identity operator from E into F is of stable type 2. Also we can consider R and U as measures on E. Therefore to apply Theorem 4.2 we need only show

(4.7) $$R\{x: \||x\||_{\tau'} > 1\} < \infty$$

and

(4.8) $$\int_{\{x: \||x\||_{\tau'} \leq 1\}} \||x\||^2_{\tau'} R(dx) < \infty.$$

Using (4.4) we see that (4.6) is equivalent to

(4.9) $$C \iint_{\{x: \||x\||_{\tau'} > 1\}} \frac{dr}{r^{1+p}} U(du).$$

We use the fact that $r = \|x\|_\infty$ and $u = x/\|x\|_\infty$, $x \neq 0$, to see that $\||x\||_{\tau'} > 1$ is equivalent to $\||\frac{x}{\|x\|_\infty}\||_{\tau'} > \frac{1}{\|x\|_\infty}$ or $\|x\|_\infty = r > 1/\||u\||_{\tau'}$. Therefore (4.9) is equal to

$$C \iint_{1/\||u\||_{\tau'}} \frac{dr}{r^{1+p}} U(du)$$

$$= \frac{C}{p} \int \||u\||^p_{\tau'} U(du)$$

which is finite by (4.6). We write (4.8) as

$$C \iint_{\{x: \||x\||_{\tau'} \leq 1\}} r^2 \||u\||^2_{\tau'} \frac{dr}{r^{1+p}} U(du).$$

By the same reasoning as above this is equal to

$$C \iint^{1/\||u\||_{\tau'}} r^2 \||u\||^2_{\tau'} \frac{dr}{r^{1+p}} U(du)$$

$$= \frac{C}{2-p} \int \||u\||^p_{\tau'} U(du)$$

and this is finite as above. Since (4.7) and (4.8) imply (4.5) Theorem 4.3 now follows from Theorem 4.2.

We now show that Theorems 4.1 and 4.3 are quite sharp in certain special cases. To see this we apply Theorem 4.1 to the stationary stable processes considered in Section 3. Consider $\int x\, dM$ as given in the statement of Theorem 4.1 along with the control measure m of M. Assume that the support of m consists of the functions $\{e^{it\lambda}, -\infty < \lambda < \infty\}$ and define m' by

$$m'(\lambda_2)-m'(\lambda_1) = m\{e^{it\lambda}, \lambda_1 < \lambda \leq \lambda_2\}.$$

Then if M' is the independently scattered random stable measure of exponent p with control measure m' we have

$$\int_F x\, dM = \int_{-\infty}^{\infty} e^{it\lambda} M'(d\lambda), \quad t\in[-1/2,1/2]. \tag{4.10}$$

Now let $\tau(t,t+h) = \tau(h)$ be a continuous symmetric pseudo-metric on $[-1/2,1/2]$ (i.e. $\tau(h) = \tau(-h)$) such that

$$\tau(h) \uparrow \quad h\in[0,1] \tag{4.11}$$

and

$$h/\tau(h) \uparrow \quad h\in[0,1]. \tag{4.12}$$

It is easy to check that

$$|||\, e^{it\lambda} |||_\tau \leq 1 + 3/\tau(1/\lambda).$$

Therefore by Theorem 4.1

$$\int_{-\infty}^{\infty} |\tau(1/\lambda)|^{-p} m'(d\lambda) < \infty \tag{4.13}$$

is a sufficient condition for the processes in (4.10) to have continuous sample paths a.s.

For τ satisfying (4.11) one can see from (3.20) that (4.1) is equivalent to

(4.14) $$\int_0^{\delta'} \frac{\tau(h)}{h(\log 1/h)^{1/2}} dh < \infty$$

for some $\delta' > 0$. The function

$$\tau(h) = (\log 10/h)^{-1/2}(\log\log 10/h)^{-(1+\delta)}, \quad \delta > 0$$

satisfies (4.14). Now, suppose that m' is concentrated on the integers so that (4.10) is a random Fourier series of the type given in (3.21). We then have that

(4.15) $$\sum_{k\geq 10} |a_k|^p (\log k)^{p/2}(\log\log k)^{p(1+\delta)} < \infty$$

is a sufficient condition for the uniform convergence a.s. of these series. In the case p = 2 this is a stronger result than the original Paley-Zygmund condition which is $\Sigma|a_k|^2(\log k)^{1+\varepsilon} < \infty$, for some $\varepsilon > 0$.

Let $a_k = k^{-1/p}(\log k)^{-\beta}$ in (3.21). Then by (4.15) the series (3.21) converges uniformly a.s. for $\beta > 1/p + 1/2$. By other considerations we know that (3.21) diverges for $\beta \leq 1$ $(1 < p \leq 2)$. These examples show that Theorem 4.1 is quite sharp in certain cases. However we note that (4.15) implies (3:22) but not conversely so Theorem 4.1 is not best possible for this class of examples.

References

1. Araujo, A. and Giné, E. The Central Limit Theorem for Real and Banach Valued Random Variables (1978), manuscript.

2. Bretagnolle, J., Dacunha-Castelle, D. and Krivine, J. L., Lois stable et espaces L^p, Ann. Inst. H. Poincare 2B (1966), 231-259.

3. Boulicaut, P., Fonctions de Young et continuité des trajectoires d'une function aléatoire, Ann. Inst. Fourier, 2 (1974), 27-48.

4. Fernique, X., Continuite et theoreme central limite pour les transformees de Fourier des measures aleatoire du second ordre, Z. Warscheinlichkeitsth., 42 (1978), 57-68.

5. Gikhman, I. I. and Skorohod, A. V., Introduction to the Theory of Stochastic Processes, (1969), W. B. Saunders, Philadelphia.

7. Jain, N. C. and Marcus, M. B. Continuity of subgaussian processes, Advances in Prob., 4 (1978), M. Decker, N. Y.

8. Marcus, M. B. and Woyczynski, W. A., Stable measures and central limit theorems in spaces of stable type, Trans. Amer. Math. Soc., to appear.

9. Marcus, M. B. and Pisier, G., Necessary and sufficient conditions for the uniform convergence of random trigonometric series, Lecture Note Series No. 50 (1978), Arhus University, Denmark.

10. Marcus, M. B. and Pisier, G., Random Fourier series on locally compact Abelian groups, Lecture Notes in Math. (Strasbourg seminar 1977-78).

11. Woyczynski, W. A., Geometry and martingales in Banach spaces, Part II: Independent increments, Probability on Banach Space, Advances in Prob. 4 (1978), M. Decker, N. Y.

12. Zinn, J., A note on the central limit theorem in Banach spaces, Ann. Prob. (1977), 283-286.

CONDITIONS FOR ABSOLUTE CONTINUITY

Charles R. Baker*

Department of Statistics
University of North Carolina
Chapel Hill, North Carolina 27514

Introduction

Several methods are available for determining absolute continuity between two artibtrary probability measures on a real separable Banach space. These methods generally have a common first step, which requires calculation of a sequence of Radon-Nikodym derivatives. It would be desirable to have general conditions for absolute continuity which do not require these calculations. In particular, it should be possible to state conditions involving only the two characteristic functions. We present here some results in this direction. We also investigate connections between weak convergence and absolute continuity, and the general applicability of conditions that are necessary and sufficient for absolute continuity when one begins with the procedure of calculating a sequence of Radon-Nikodym derivatives. Finally, some specialized results for Gaussian measures are given.

We assume throughout that $\mathcal{B}$ is a real separable Banach space, with norm $||\cdot||$ and Borel σ-field $B[\mathcal{B}]$, and that ν and μ are probability measures on $B[\mathcal{B}]$. All measures considered here are countably additive. We consider conditions to determine if ν is absolutely continuous with respect to μ ($\nu << \mu$). If f is a real-valued function in $L_1[\mu]$, we denote by $T_\mu f$ its Fourier transform with respect to μ; $[T_\mu f](x) = \int_{\mathcal{B}} e^{ix(y)} f(y)d\mu(y)$, for x in $\mathcal{B}^*$. T_μ is a 1:1 continuous linear map of $L_1[\mu]$ into $C[\mathcal{B}^*]$ (the space of bounded continuous functions on $\mathcal{B}^*$). $\hat{\nu}$ and $\hat{\mu}$ will denote the characteristic functions of ν and μ. An obvious necessary and sufficient condition for $\nu << \mu$ is that $\hat{\nu}$ belong to range(T_μ). If (λ_n) is a sequence of finite positive measures on $B[\mathcal{B}]$, then $\lambda_n \overset{w}{\Rightarrow} \nu$ denotes weak convergence of (λ_n) to ν in the Alexandroff-Prohorov sense [1]. A family Λ of measures on $B[\mathcal{B}]$ is said to be uniformly Radon if the following condition is satisfied: For every Borel set A, and

*Research partially supported by ONR contract N00014-75-C-0491.

any $\varepsilon > 0$, there exists a compact set $K \subset A$ (depending on A and ε) such that $|\lambda|(A-K) < \varepsilon$ for all $\lambda \in \Lambda$, where $|\lambda|$ denotes the total variation of λ. The family is uniformly countably additive if when $\{E_m, m \geq 1\}$ is a decreasing family of Borel sets, $\bigcap_m E_m = \phi$, then $|\lambda|(E_m) \to 0$ uniformly in λ. Λ is uniformly dominated by μ if for any $\varepsilon > 0$ there exists $\delta > 0$ such that $\mu(A) < \delta \Rightarrow |\lambda|(A) < \varepsilon$, all $\lambda \in \Lambda$. If Λ is bounded (in variation norm), then Λ is uniformly dominated by μ if and only if $\lambda << \mu$ for all λ in Λ ($\Lambda << \mu$) and $\{d\lambda/d\mu: \lambda \in \Lambda\}$ is uniformly integrable with respect to μ. We assume throughout that $\mathcal{B}$ is infinite-dimensional.

Conditions for Absolute Continuity Determined by Sequences

A general method for determining if absolute continuity holds is based on the following well-known result.

Lemma 1. Let $\{F_n, n \geq 1\}$ be an increasing family of sub-σ-fields of $B[\mathcal{B}]$ such that $B[\mathcal{B}]$ is the smallest σ-field containing $\bigcup_{n\geq 1} F_n$. Let ν_n (resp., μ_n) be the restriction of ν (resp., μ) to F_n. Then $\nu << \mu$ if and only if

(a) $\nu_n << \mu_n$ for all $n \geq 1$

(b) $\{d\nu_n/d\mu_n, n \geq 1\}$ is uniformly integrable with respect to μ.

Proof: (a) is obviously necessary; necessity of (b) follows from the fact that $d\nu_n/d\mu_n = E[d\nu/d\mu|F_n]$ if $\nu << \mu$. Sufficiency can be proved by utilizing the fact that $\{d\nu_n/d\mu_n, F_n; n \geq 1\}$ is a positive martingale with respect to μ (see, e.g., [2], pp. 441-443); it also follows from Theorem 1 below. □

One can easily show that if $f_n \equiv d\nu_n/d\mu_n$ for ν_n and μ_n as in Lemma 1, then $[T_\mu f_n](x) \to \hat{\nu}(x)$ uniformly on $\mathcal{B}^*$ is a necessary condition for $\nu << \mu$, as is $\lambda_n \overset{w}{\Rightarrow} \nu$, $d\lambda_n \equiv f_n d\mu$.

In order to apply Lemma 1, it is necessary to verify that $\nu_n << \mu_n$ for $n \geq 1$, and to compute the family of Radon-Nikodym derivatives $\{d\nu_n/d\mu_n, n \geq 1\}$. Here we shall investigate conditions for absolute continuity in terms of sequences $\{f_n, n \geq 1\} \subset L_1[\mu]$ which have a more general form; in particular, more general sequences need not have the martingale property enjoyed by $\{d\nu_n/\mu_n, n \geq 1\}$. Of course, any such sequence (f_n) must have some property linking it to ν; we shall impose the reasonably weak requirement that $[T_\mu f_n](x) \to \hat{\nu}(x)$ for all x in B^*.

Parts (2) and (3) of the next lemma have been previously proved for Hausdorff spaces (along with a number of other interesting results) by Gänssler [3]. Part (2) is an extension of a result due to Grothendieck [4] for locally compact spaces. We include here an independent proof, using aspects of weak convergence.

Lemma 2. Suppose that Λ is a bounded family of measures on $\mathcal{B}[B]$.

(1) If (λ_n) is a positive uniformly Radon sequence in Λ, $\hat{\lambda}_n \to \hat{\nu}$ on B^*, and $\lambda_n << \mu$ for all $n \geq 1$, then $\nu << \mu$.

(2) Λ is uniformly Radon if and only if Λ is uniformly dominated by some positive finite measure τ on $\mathcal{B}[B]$.

(3) If $\Lambda << \mu$, then Λ is uniformly dominated by μ if and only if Λ is uniformly Radon.

Proof: (1) $\hat{\lambda}_n \to \hat{\nu}$ on B^* and (λ_n) uniformly Radon imply $\lambda_n \overset{W}{\Rightarrow} \nu$. If $\nu << \mu$ is not true, there exists a Borel set A such that $\nu(A) > 0$ and $\mu(A) = 0$. (λ_n) uniformly Radon and $\lambda_n(A) = 0$ for $n \geq 1$ imply that there exists an open set $\mathcal{O} \supset A$ such that $\lambda_n(\mathcal{O}\text{-}A) = \lambda_n(\mathcal{O}) < \nu(A)$ for all $n \geq 1$. Thus $\underline{\lim}\, \lambda_n(\mathcal{O}) < \nu(A) \leq \nu(\mathcal{O})$, a contradiction of $\lambda_n \overset{W}{\Rightarrow} \nu$.

(2) It is clear that uniform dominance by any finite positive τ implies Λ is uniformly Radon, since such a τ is Radon. For the converse, define Λ' as the set of positive measures $\{|\lambda| : \lambda \in \Lambda\}$. Let (λ_n) be any sequence in Λ, and define ω by $\omega = \sum_{n\geq 1} 2^{-n} |\lambda_n|/|\lambda_n|(B)$; $|\lambda_n| << \omega$ for $n \geq 1$. Λ' uniformly Radon and bounded implies Λ' is relatively weakly compact, and so there is a subsequence (λ_{n_k}) and a finite positive measure θ such that $|\lambda_{n_k}| \overset{W}{\Rightarrow} \theta$. By part (1), $\theta << \omega$. Now suppose (λ_{n_k}) is not uniformly dominated by ω. Then there exists $\varepsilon > 0$ such that for any $\delta > 0$ there exists A_δ with $\omega(A_\delta) < \delta$ and $\sup_k |\lambda_{n_k}|(A_\delta) > \varepsilon$. Since $\theta << \omega$, there exists $\delta_j \downarrow 0$ and a Borel sequence (A_j) such that $\omega(A_j) < \delta_j$, $\theta(A_j) < \frac{1}{2}\varepsilon j^{-2}$, $|\lambda_{n_{k_j}}|(A_j) \geq \varepsilon$, all $j \geq 1$, for $\{\lambda_{n_{k_j}}, j \geq 1\} \subset \{\lambda_{n_k}, k \geq 1\}$. Let $A = \bigcup_j A_j$. $\theta(A) < \frac{1}{2}\varepsilon$, $|\lambda_{n_{k_j}}|(A) \geq \varepsilon$, $j \geq 1$. Since each $|\lambda_{n_k}| << \mu$, we can assume that $n_{k_j} < n_{k_{j+1}}$ for $j \geq 1$. By the uniform

Radon property, there exists a compact $K \subset A$ such that $|\lambda_{n_{k_j}}|(A-K) < \varepsilon/4$ for $j \geq 1$. Thus $|\lambda_{n_{k_j}}|(K) > |\lambda_{n_{k_j}}|(A) - \varepsilon/4 \geq 3\varepsilon/4$, $j \geq 1$. Hence $\overline{\lim}|\lambda_{n_{k_j}}|(K) \geq 3\varepsilon/4 \geq \theta(A) \geq \theta(K)$, which contradicts the fact that $|\lambda_{n_{k_j}}| \xrightarrow{w} \theta$. Thus, every infinite sequence in Λ' contains an infinite subsequence that is uniformly dominated by a positive finite measure ω (depending on the subsequence). This is equivalent to every infinite sequence in Λ' containing an infinite uniformly countably additive subsequence ([5], V.9.1, V.9.2, V.9.3), which is equivalent to uniform countable additivity of Λ'. Λ' uniformly countably additive and bounded implies Λ' is uniformly dominated by some positive finite measure τ [5].

(3) This follows from part (2) and the fact (due to Dubrovskii; see [5], p. 389) that Λ uniformly dominated by some positive measure τ implies Λ uniformly dominated by every positive measure ρ such that $\Lambda << \rho$. □

Remark. The proof of Lemma 2 holds without change when B is only assumed to be a complete separable metric space.

The following result collects a number of conditions for $\nu << \mu$.

Theorem 1: Suppose that (f_n) is a sequence in $L_1[\mu]$ such that $T_\mu f_n \to \hat{\nu}$ on B^*. Define the sequence of measures (λ_n) by $d\lambda_n = |f_n|d\mu$. Consider the following conditions:

(1) $\{f_n, n \geq 1\}$ is uniformly integrable with respect to μ

(2) $\{f_n, n \geq 1\}$ is relatively compact in the $\sigma(L_1, L_\infty)$ topology

(3) $\{\lambda_n, n \geq 1\}$ is uniformly countably additive and bounded

(4) There exists a non-negative non-decreasing function ϕ on $\mathbb{R}^1_+$ such that $\lim_{t\to\infty} \phi(t)/t = \infty$ and $\sup_n \int_B \phi[|f_n|]d\mu < \infty$

(5) $\{\lambda_n, n \geq 1\}$ is uniformly Radon and bounded

(6) (f_n) converges in $L_1[\mu]$

(7) For some f in $L_1[\mu]$, $||f_n||_1 \to ||f||_1$ and $f_n \to f$ a.e. $d\mu$.

Each of conditions (1)-(7) implies $\nu << \mu$. The converse is false; one can have $\nu << \mu$ and none of conditions (1)-(7) satisfied. (λ_n) relatively weakly compact (r.w.c.) is necessary for any of conditions (1)-(7) to hold; (λ_n) r.w.c. is neither necessary nor sufficient for $\nu << \mu$. If there exists an a.e. $d\mu$ finite measurable function f such that $f_n \to f$ a.e. $d\mu$, then conditions (1)-(7) are equivalent.

<u>Proof</u>: It is well-known ([5], [6]) that (1) <=> (2) <=> (3) <=> (4), and that (7) => (6) => (2). By Lemma 2, (1) <=> (5). $f_{n_k} \to f$ in $\sigma(L_1, L_\infty)$ implies $T_\mu f_{n_k} \to T_\mu f$ on B^*, so (2) => $\nu << \mu$, and thus conditions (1)-(7) are sufficient for $\nu << \mu$. If $f_n \to f$ a.e. $d\mu$, f measurable and finite a.e. $d\mu$, then (1) => (7) [6], so that conditions (1)-(7) are equivalent. (λ_n) is relatively weakly compact if (λ_n) is bounded and uniformly tight [1]; both these conditions are necessary for (5) to hold. Thus, (λ_n) r.w.c. is necessary for any of (1)-(7) to hold. The fact that (λ_n) r.w.c. is neither necessary nor sufficient for $\nu << \mu$ will be shown below (Prop. 1). □

Suppose (F_n) is an increasing sequence of sub-σ-fields of $B[B]$, such that $B[B] = \sigma[\bigcup_n F_n]$. If $\nu << \mu$ and $f_n \equiv E[d\nu/d\mu | F_n]$, then $\{f_n, F_n: n \geq 1\}$ is a positive μ-martingale and $f_n \to d\nu/d\mu$ a.e. $d\mu$. Thus, conditions (1)-(7) are necessary for $\nu << \mu$ in this case (the framework of Lemma 1). If $\nu << \mu$ on each F_n, and $f_n \equiv d\nu|_{F_n}/d\mu|_{F_n}$, then condition (4) of Theorem 1 yields Hajek's well-known divergence criterion [7] by choosing $\phi(t) = t \log t$ for $t > 0$, $\phi(0) = 0$. Satisfaction of condition (4) for this specific ϕ is also *necessary* for $\nu << \mu$ if both ν and μ are Gaussian.

As noted, in the framework of Lemma 1 both of the following conditions are necessary for $\nu << \mu$: (1) $T_\mu f_n \to \hat{\nu}$ uniformly on B^*, and (2) $\lambda_n \overset{w}{\Rightarrow} \nu$, where $d\lambda_n = [d\nu_n/d\mu_n]d\mu$. We proceed to show that neither of these conditions is necessary for $\nu << \mu$ in the general case ($T_\mu f_n \to \hat{\nu}$ on B^*). $\hat{\nu} << \hat{\mu}$ means that for every $\varepsilon > 0$ there exists $\delta > 0$ such that $|1-\hat{\nu}(x)| < \varepsilon$ when $|1-\hat{\mu}(x)| < \delta$; for a family Λ of probability measures, $\hat{\Lambda} << \hat{\mu}$ means that $\hat{\lambda} << \hat{\mu}$ for each $\lambda \in \Lambda$. $\hat{\nu} << \hat{\mu}$ is necessary, but in general not sufficient, for $\nu << \mu$ [8].

<u>Lemma 3</u>. Suppose B is a Hilbert space. If Λ is a family of probability measures on $B[B]$, then each of the following is sufficient for relative weak compactness of Λ:

(1) For each $\varepsilon > 0$ and $\lambda \in \Lambda$, there exists a covariance operator $R_{\lambda,\varepsilon}$ such that $\langle R_{\lambda,\varepsilon}x, x\rangle \leq 1$ implies $1\text{-Re } \hat{\lambda}(x) < \varepsilon$, and $R_{\lambda,\varepsilon} \leq R_\varepsilon$ for all $\lambda \in \Lambda$, where R_ε is a covariance operator.

(2) $\hat{\Lambda} << \hat{\tau}$ uniformly for some probability measure τ.

(3) Each infinite sequence in Λ contains an infinite subsequence (λ_n) such that $\hat{\lambda}_n$ converges uniformly to some $\hat{\omega}$ (ω a probability measure depending on the subsequence).

Proof: (1) This follows from Theorem 1.14 of [1].

(2) For each $\delta > 0$ there exists a covariance operator R_δ such that $<R_\delta x,x> \leq 1$ implies $1\text{-Re } \hat{\tau}(x) < \delta$ [1]. For a given $\varepsilon > 0$, using $\hat{\Lambda} << \hat{\tau}$ uniformly, choose δ so that $1\text{-Re } \hat{\tau}(x) < \delta$ implies $1\text{-Re } \hat{\lambda}(x) < \varepsilon$, all $\lambda \in \Lambda$. Then $<R_\delta x,x> \leq 1$ implies $1\text{-Re } \hat{\lambda}(x) < \varepsilon$ for all λ in Λ, and Λ is relatively weakly compact from (1).

(3) $|1-\hat{\lambda}_n(x)| \leq |1-\hat{\omega}(x)| + |\hat{\lambda}_n(x) - \hat{\omega}(x)|$. Fix $\varepsilon > 0$. There exists a covariance operator R_ε such that $<R_\varepsilon x,x> \leq 1$ implies $|1-\hat{\omega}(x)| < \varepsilon/2$. Moreover, by the uniform convergence there exists N_ε such that $n > N_\varepsilon$ implies $|\hat{\lambda}_n(x) - \hat{\omega}(x)| < \varepsilon/2$, for all x in B^*. Let $\tau \equiv \frac{1}{N_\varepsilon} \sum_1^{N_\varepsilon} \lambda_n$. Then $\lambda_n << \tau$, $1 \leq n \leq N_\varepsilon$. Choose $\delta_n > 0$ so that $|1-\hat{\tau}(x)| < \delta_n \Rightarrow |1-\hat{\lambda}_n(x)| < \varepsilon/2$, $n = 1,\ldots,N_\varepsilon$, and set $\delta_0 = \inf\{\delta_1,\ldots,\delta_{N_\varepsilon}\}$. Let S_ε be a covariance operator such that $<S_\varepsilon x,x> \leq 1$ implies $|1-\hat{\tau}(x)| < \delta_0$. Then $<(R_\varepsilon + S_\varepsilon)x, x> \leq 1$ implies $|1-\hat{\lambda}_n(x)| < \varepsilon$ for all $n \geq 1$, so that (λ_n) is relatively weakly compact, and thus $\lambda_n \overset{w}{\Rightarrow} \omega$. □

Prop. 1: (1) Let (λ_n) be a sequence of probability measures on $B[B]$. If $\lambda_n << \mu$ for $n \geq 1$, $\hat{\lambda}_n(x) \to \hat{\nu}(x)$ for all x in B^*, and $\lambda << \mu$, it is not necessary that $\lambda_n \overset{w}{\Rightarrow} \nu$.

(2) For every zero-mean Gaussian measure μ on $B[B]$ with infinite-dimensional support there exists a Gaussian family $\{\nu,\lambda_n, n\geq 1\}$ on $B[B]$ with $\hat{\lambda}_n \to \hat{\nu}$ on B^*, $\nu \perp \mu$, and $\{\hat{\lambda}_n, n \geq 1\} << \hat{\mu}$ uniformly. There is also a Gaussian family $\{\lambda_n', n \geq 1\}$ such that $\lambda_n' \perp \mu$ for $n \geq 1$, while $\hat{\lambda}_n' \to \hat{\mu}$ on B^* and $\{\hat{\lambda}_n', n \geq 1\} << \hat{\mu}$ uniformly.

Proof: (1) Pick any sequence of probability measures (λ_n) such that $\hat{\lambda}_n \to \hat{\nu}$ on B^*, with (λ_n) not relatively weakly compact. The counterexample is obtained by considering the measure $\mu = \sum_{n\geq 1} 2^{-(n+1)}\lambda_n + \frac{1}{2}\nu$.

(2) To prove the first assertion, suppose that $\hat{\mu}(x) = \exp\{-\frac{1}{2} R(x,x)\}$, R the covariance of μ. Define $\hat{\nu}(x) = \exp\{-R(x,x)\}$. Let $W: B \to L_2[0,1]$ be 1:1, linear,

and continuous (W can be taken as the composition of a linear isometry of $\mathcal{B}$ into $C[0,1]$ and the inclusion map of $C[0,1]$ into $L_2[0,1]$). Let S be the covariance operator of the Gaussian measure $\mu \circ W^{-1}$ on the Borel σ-field of $L_2[0,1]$. The Gaussian measure $\nu \circ W^{-1}$ has covariance 2S; from well-known results, $\nu \circ W^{-1} \perp \mu \circ W^{-1}$, so that $\nu \perp \mu$. Let λ_n be the zero-mean Gaussian measure on $B[\mathcal{B}]$ with covariance $[(4n-1)/(2n)]R$. $\hat{\lambda}_n(x) \to \exp\{-R(x,x)\} = \hat{\nu}(x)$ for all x in $\mathcal{B}^*$. $\hat{\lambda}_n \geq [\hat{\mu}]^2$ on $\mathcal{B}^*$, all $n \geq 1$, so $\{\hat{\lambda}_n, n \geq 1\} << \hat{\mu}$ uniformly. This completes the proof of the first assertion of (2); the second assertion can be proved by defining $\hat{\lambda}'_n(x) = \exp\{-\frac{1}{2}(1 + 1/n)R(x,x)\}$. □

Remark. In proving the first assertion in part (2) of Prop. 1, the sequence (λ_n) is such that $\lambda_n \perp \mu$ for $n \geq 1$. However, this is not a critical factor; if $\mathcal{B}$ is a Hilbert space, one can show that the assertion is true for a sequence (λ_n) such that $\lambda_n << \mu$ and $\mu << \lambda_n$ for all $n \geq 1$. Let μ be zero-mean Gaussian with covariance operator $R = \sum_k \beta_k e_k \otimes e_k$, where $\{e_k, k \geq 1\}$ is orthonormal, $\beta_k > 0$ for $k \geq 1$, and $\sum_k \beta_k < \infty$. Let ν be zero-mean Gaussian with covariance operator 2R, and take λ_n to be zero-mean Gaussian with covariance $R + \sum_1^n \beta_k e_k \otimes e_k$. The first assertion of (2) is easily seen to be satisfied; further, λ_n and μ are mutually absolutely continuous for all $n \geq 1$.

From Prop. 1, we conclude that $\lambda_n \overset{w}{\Rightarrow} \nu$ is not always necessary for $\nu << \mu$, when $\lambda_n << \mu$ for $n \geq 1$; if $\mathcal{B}$ is a Hilbert space, then this result and Lemma 3 imply that neither $\{\hat{\lambda}_n, n \geq 1\} << \hat{\mu}$ uniformly nor $\hat{\lambda}_n \to \hat{\nu}$ uniformly on $\mathcal{B}^*$ is a necessary condition for $\nu << \mu$. Further, $\{\hat{\lambda}_n, n \geq 1\} << \hat{\mu}$ uniformly (and thus $\lambda_n \overset{w}{\Rightarrow} \nu$) is not sufficient for $\nu << \mu$ when $\hat{\lambda}_n \to \hat{\nu}$ on $\mathcal{B}^*$, $\lambda_n << \mu$ for $n \geq 1$, if $\mathcal{B}$ is Hilbert.

Finally, we remark that since ν and μ are Radon, it will always be true that $\nu << \mu$ if and only if: K compact and $\mu(K) = 0$ imply $\nu(K) = 0$. However, if $\lambda_n \overset{w}{\Rightarrow} \nu$ and $\lambda_n << \mu$ for all $n \geq 1$, this can be strengthened: $\nu << \mu$ if and only if K compact and $\mu(\partial K) = 0$ imply $\nu(\partial K) = 0$ (where ∂K is the boundary of K). The sufficiency of this condition follows from $\underline{\lim}\, \lambda_n(\mathcal{O}) \geq \nu(\mathcal{O})$ for every open set $\mathcal{O}$ (as $\lambda_n \overset{w}{\Rightarrow} \nu$) and the fact that ν is Radon.

Topological Conditions for Absolute Continuity

A problem of major interest is to characterize all probability measures ν that are absolutely continuous with respect to a fixed Gaussian measure μ. This problem arises in various applications: Information Theory, filtering, and detection of signals in noise. Here we give necessary and sufficient conditions for $\nu << \mu$ for a class of measures μ that includes all Gaussian measures. These conditions are stated in terms of the characteristic functional of ν and a topology determined by T_μ. The complete proof (and applications) of these results will appear elsewhere [9].

If X is a linear space and Γ a total linear space of linear functionals on X, the Γ topology on X is given by a neighborhood base at zero having sets of the form $\bigcap_1^N \{x: |\gamma_i(x)| < \delta \}$, where $N \geq 1$, $\delta > 0$, and $\gamma_1,\ldots,\gamma_N$ range over Γ. K will denote the linear space consisting of range(T_μ) and all bounded continuous functions on $\mathcal{B}^*$ that are pointwise limits of sequences in range(T_μ). Π_1' is the set of all complex-valued finite measures on $\mathcal{B}^*$ whose support is a finite subset of $\mathcal{B}^*$. We define an equivalence relation on Π_1' by $\pi_1 \sim \pi_2$ if $\pi_1(T_\mu f) = \pi_2(T_\mu f)$ for all f in $L_1[\mu]$. Π_1 is the collection of all equivalence classes from Π_1'. In the case $\mathcal{B}$ is a Hilbert space, H, let Π_2 be the set of all complex-valued finite measures on H that are absolutely continuous with respect to μ. Π_1 and Π_2 are linear spaces over the complex numbers. By a linear variety in $\mathcal{B}$ we mean a set of the form S + m, where $m \in \mathcal{B}$ and S is a closed linear manifold.

Theorem 2: (1) If for any x in $\mathcal{B}^*$ there exists f in $L_1[\mu]$ such that $[T_\mu f](x) \neq 0$, then $\nu << \mu$ if and only if $G: \pi \to \pi(\hat{\nu})$ is continuous on Π_1 in the range(T_μ) topology.

(2) If $\mathcal{B}$ is a Hilbert space and support (μ) is a linear variety, then $\nu << \mu$ if and only if $\hat{\nu} \in K$ and $G: \pi \to \pi(\hat{\nu})$ is continuous on Π_2 in the range(T_μ) topology.

Outline of proof: One first shows that range(T_μ), viewed as a subset of the algebraic dual of Π_i ($[T_\mu f](\pi) \equiv \pi[T_\mu f]$), is total over Π_i $(i = 1,2)$. In the case of Π_1, this is immediate. For Π_2, this is satisfied if T_μ is a 1:1 map of $L_1[\mu]$

into $L_1[\mu]$, and this can be proved by using the continuity of $T_\mu f$ and the structure of $\mathrm{supp}(\mu)$. Range(T_μ) total over Π_i implies that $G\colon \pi \to \pi(\hat{\nu})$ is continuous on Π_i in the range(T_μ) topology if and only if $\pi(\hat{\nu}) = \pi(T_\mu f)$ for some f in $L_1[\mu]$, all π in Π_i. If $\nu << \mu$, this is satisfied by $f = d\nu/d\mu$. If $\pi(\hat{\nu}) = \pi(T_\mu f)$ for all π in Π_1, the assumption in (1) implies that π_x ($\pi_x[g] \equiv g[x]$) is not (equivalent to) the zero element in Π_1, for any x in B^*, so that $\hat{\nu} = T_\mu f$ on B^*. Finally, if $\pi(\hat{\nu}) = \pi(T_\mu f)$ for all π in Π_2, then $\hat{\nu} = T_\mu f$ on B^* is proved using the continuity of $T_\mu f$ on B^* and the assumption that $\hat{\nu} \in K$. □

Special Results for Gaussian Measures

Suppose μ_{XY} is a zero-mean Gaussian measure on $B[B] \times B[B]$, with marginals μ_X and μ_Y; e.g., $\mu_X(A) = \mu_{XY}(A \times B)$. $\mu_X \otimes \mu_Y$ is the usual product measure. Let H_X (resp., H_Y) be the set of all elements m in B such that the translate of μ_X (resp., μ_Y) by m is absolutely continuous with respect to μ_X (resp., μ_Y). Let $\mu_{X+Y} \equiv \mu_{XY} \circ f^{-1}$, f the addition map; define H_{X+Y} in the same manner as H_X and H_Y. Let $\mu_{X,X+Y}(A) \equiv \mu_{XY}\{(x,y)\colon (x, x+y) \in A\}$. The "average mutual information" $I(\mu_{XY})$ in the measure μ_{XY} is defined as

$$I(\mu_{XY}) = \int_{B\times B} \log \left[\frac{d\mu_{XY}}{d\mu_X \otimes \mu_Y}\right] d\mu_{XY} \qquad \text{if } \mu_{XY} << \mu_X \otimes \mu_Y$$

$$= +\infty \qquad \text{otherwise} .$$

$I(\mu_{X,X+Y})$ is defined in a similar manner.

Suppose that B is a Hilbert space; then $H_X = \mathrm{range}(R_X^{1/2})$, R_X the covariance operator of μ_X; similarly for H_Y, H_{X+Y}, $R_Y^{1/2}$ and $R_{X+Y}^{1/2}$. In this case, Pitcher has proved the following result.

Lemma 4 [10]. If μ_{XY} is Gaussian, $\mu_{XY} = \mu_X \otimes \mu_Y$, and $\mu_{X+Y} << \mu_Y$, then the following are equivalent:

(1) $I(\mu_{X,X+Y}) < \infty$;

(2) $\mu_X[\mathrm{range}(R_Y^{1/2})] = 1$;

(3) μ_{X+Y} and μ_Y are *strongly* equivalent; that is, $R_{X+Y} = R_Y^{1/2}(I+S)R_Y^{1/2}$ for S *trace-class* (and -1 not in $\sigma(S)$).

Strong equivalence is a term coined by Hajek [11]; the existence of strong equivalence provides some advantages in computing the Radon-Nikodym derivative $d\mu_{X+Y}/d\mu_Y$.

The assumption that $\mu_{X+Y} << \mu_Y$ was made by Pitcher in order to utilize the fact (in his proof) that $R_X = R_Y^{\frac{1}{2}} S R_Y^{\frac{1}{2}}$ for S compact. However, the following generalization can be made.

Lemma 5 [12]. Suppose μ_{XY} is Gaussian, and that $I(\mu_{XY}) < \infty$. Then $\mu_{X+Y} << \mu_Y$ and properties (1)-(3) of Lemma 4 are equivalent.

Remark. If $\mu_{XY} = \mu_X \otimes \mu_Y$, as in Pitcher's result, then $I(\mu_{XY}) = 0$.

Proof: The complete proof will appear in [12]. We sketch here the proof that $\mu_X[\text{range}(R_Y^{\frac{1}{2}})] = 1$ implies $I(\mu_{X,X+Y}) < \infty$.

Let R_{XY} be the covariance operator of μ_{XY}; $R_{XY} = R_X^{\frac{1}{2}} V R_Y^{\frac{1}{2}}$, where $||V|| \leq 1$, $P_X V = V P_Y = V$, P_X(resp, P_y) the projection operator with range equal to $\overline{\text{range}(R_X)}$ (resp., $\overline{\text{range}(R_Y)}$). Moreover, $I(\mu_{XY}) < \infty$ (for μ_{XY} Gaussian) if and only if $||V|| < 1$ and V is Hilbert-Schmidt. Similarly, $R_{X,X+Y} = R_X^{\frac{1}{2}} U R_{X+Y}^{\frac{1}{2}}$, $||U|| \leq 1$, $P_X U = U P_{X+Y} = U$. We show that $\mu_X[\text{range}(R_Y^{\frac{1}{2}})] = 1$ implies $||U|| < 1$ and U Hilbert-Schmidt. $I(\mu_{XY}) < \infty$ can be shown to imply $R_X^{\frac{1}{2}} = R_Y^{\frac{1}{2}} P$ for P bounded, and also $R_{X+Y}^{\frac{1}{2}} = R_Y^{\frac{1}{2}} B$ for B bounded with bounded inverse. $\mu_X[\text{range}(R_Y^{\frac{1}{2}})] = 1$ and μ_X Gaussian imply that P is Hilbert-Schmidt. Now $R_{X+Y} = R_Y^{\frac{1}{2}} BB^* R_Y^{\frac{1}{2}} = R_X + R_{XY} + R_{YX} + R_Y = R_Y^{\frac{1}{2}}(PP^* + PV + V^*P^* + I)R_Y^{\frac{1}{2}}$. $R_{X,X+Y} = R_X + R_{XY} = R_X^{\frac{1}{2}} P^* R_Y^{\frac{1}{2}} + R_X^{\frac{1}{2}} V R_Y^{\frac{1}{2}} = R_X^{\frac{1}{2}} U R_{X+Y}^{\frac{1}{2}} = R_X^{\frac{1}{2}} U B^* R_Y^{\frac{1}{2}}$, so that $P^* + V = UB^*$. Since P and V are Hilbert-Schmidt, U is Hilbert-Schmidt. Moreover, $BB^* = PP^* + PV + V^*P^* + I$ and $U = (P^* + V)B^{*-1}$ imply that $U^*U = B^{-1}(P + V^*)(P^* + V)B^{*-1} = B^{-1}[BB^* - I + V^*V)B^{*-1} = I + B^{-1}[V^*V - I]B^{*-1}$. U compact implies that $||U|| = 1$ if and only if $U^*Ux = x$ for some non-zero x. From the preceding expression for U^*U, $U^*Ux = x$ if and only if $V^*V - I$ has $B^{*-1}x$ in its null space; this is impossible, since $||V|| < 1$. □

One may ask if Lemma 5 cannot be further improved; the answer is essentially negative. If the assumption that $I(\mu_{XY}) < \infty$ is omitted, one can construct for each of the conditions in Lemma 4 a Gaussian measure μ_{XY} so that this condition is satisfied, but at least one of the remaining conditions is not satisfied. If the assumption that μ_{XY} is Gaussian is omitted, one can construct a measure $\mu_{XY} = \mu_X \otimes \mu_Y$ with μ_Y Gaussian and such that $\mu_X[\text{range}(R_Y^{\frac{1}{2}})] = 1$ but μ_{X+Y} and μ_Y are not strongly equivalent (in the sense that $R_{X+Y} \neq R_Y^{\frac{1}{2}}(I + T)R_Y^{\frac{1}{2}}$ for T trace-class). If μ_Y is Gaussian, μ_X pre-Gaussian, $\mu_{XY} = \mu_X \otimes \mu_Y$, one can show that μ_{X+Y} strongly equivalent to μ_Y implies $I(\mu_{X,X+Y}) < \infty$ and $\mu_X[\text{range}(R_Y^{\frac{1}{2}})] = 1$; and that $\mu_X[\text{range}(R_Y^{\frac{1}{2}})] = 1$ implies $\mu_{X+Y} \sim \mu_Y$ and $\mu_{X,X+Y} \sim \mu_X \otimes \mu_{X+Y}$.

A result similar to Lemma 5 can be obtained for the Banach space framework. First, we recall that H_Y and H_{X+Y} can each be endowed with a Hilbert space structure in various equivalent ways (e.g., [8]); hereafter, we assume this structure.

Theorem 3. Suppose μ_{XY} is Gaussian, and that $I(\mu_{XY}) < \infty$. The following statements are then equivalent:

(1) $I(\mu_{X,X+Y}) < \infty$;

(2) $\mu_X[H_Y] = 1$;

(3) $\mu_{X+Y} << \mu_Y$ and $AA^* = I + K$ (where $A: H_Y \to H_{X+Y}$ is the inclusion map) for a trace-class operator K in H_{X+Y}.

No two of the statements (1)-(3) are equivalent if one omits the assumption that $I(\mu_{XY}) < \infty$. Moreover, if $\mu_{XY} = \mu_X \otimes \mu_Y$, the statements (1)-(3) are not equivalent if one omits the assumption that μ_X is Gaussian.

Proof: We apply a 1:1 continuous linear map $T: \mathcal{B} \to L_2[0,1]$. The induced measure $\mu_{XY} \circ W^{-1}$ is then Gaussian ($W(u,v) = (Tu,Tv)$) with marginals $\mu_X \circ T^{-1}$ and $\mu_Y \circ T^{-1}$. Since $\mu_{X,X+Y} \circ W^{-1}(A) = \mu_{XY} \circ W^{-1}\{(x,y): (x,x+y) \in A\}$, the fact that $W(u,v) = (Tu,Tv)$ with T 1:1 implies [13] $I(\mu_{XY} \circ W^{-1}) = I(\mu_{XY})$ and $I(\mu_{X,X+Y} \circ W^{-1}) = I(\mu_{X,X+Y})$. $T[H_Y] = \text{range}(R_Y^{\frac{1}{2}})$, R_Y the covariance operator of $\mu_Y \circ T^{-1}$. It is thus clear that $I(\mu_{XY}) < \infty \iff I(\mu_{XY} \circ W^{-1}) < \infty$, and that the equivalence of (1) and (2) of Theorem 3 follows from the corresponding results of Lemma 5. Property (3) of Theorem 3 can be shown to be equivalent to the strong equivalence of $\mu_{X+Y} \circ T^{-1}$ and $\mu_Y \circ T^{-1}$, using Lemma 3 of [8]. Property (3) is then equivalent to (1) and (2), from Lemma 5. The remainder of Theorem 3 follows from similar results for $L_2[0,1]$.

Acknowledgement

I wish to thank S.D. Chatterji for a discussion leading to an improved Lemma 2, and David Fremlin for telling me of Gänssler's prior work [3], while at Oberwolfach.

References

[1] Yu. V. Prohorov, Convergence of Random Processes and Limit Theorems in Probability Theory, *Theory Probability Appl.*, 1, 157-214 (1956).

[2] I.I. Gihman and A.V. Skorohod, The Theory of Stochastic Processes, I, Springer-Verlag, Berlin (1974).

[3] P. Gänssler, Compactness and sequential compactness in spaces of measures, *Z. Wahrscheinlichkeitstheorie verw. Geb.*, 17, 124-146 (1971).

[4] A. Grothendieck, Sur les applications linéaires faiblement compactes d'espaces du type C(K), *Canadian J. Math*, 5, 129-173 (1953).

[5] N. Dunford and J.T. Schwartz, Linear Operators. Part I, Interscience, New York (1958).

[6] P.A. Meyer, Probability and Potentials, Blaisdell, Waltham (1966).

[7] J. Hajek, A property of the J-divergences of marginal probability distributions, *Czech. Math. J.*, 8, 460-463 (1958).

[8] C.R. Baker, Absolute Continuity for a Class of Probability Measures, *Lecture Notes in Math.* (Vector Space Measures, I, Dublin, 1977) 644, 44-50 (1978).

[9] C.R. Baker, Characteristic Function Conditions for Absolute Continuity, to appear.

[10] T.S. Pitcher, On the sample function of processes which can be added to a Gaussian process, *Ann. Math. Stat.*, 34, 329-333 (1963).

[11] J. Hajek, On linear statistical problems in stochastic processes, *Czech. Math. J.*, 87, 404-444 (1962).

[12 C.R. Baker, Mutual information, strong equivalence, and signal sample path properties for Gaussian processes, to appear, *Information and Control*.

[13] C.R. Baker, Capacity of the Gaussian channel without feedback, *Information and Control*, 37, 70-89 (1978).

A CHARACTERIZATION OF ALMOST SURE CONVERGENCE

A. Bellow[1] and A. Dvoretzky[2]

Department of Mathematics
Northwestern University
Evanston, Illinois
U.S.A.

Institute of Mathematics
The Hebrew University
Jerusalem
Israel

The paper is divided as follows:

§1. Introduction

Throughout this paper $(\Omega,\mathcal{F},P)$ is a fixed probability space and $(\mathcal{F}_n)_{n\in N}$ an increasing sequence of sub-σ-fields of $\mathcal{F}$ (i.e., $\mathcal{F}_m \subset \mathcal{F}_n \subset \mathcal{F}$ for $m \leq n$). We denote by $\mathcal{F}_\infty$ the σ-field spanned by $\bigcup_{n\in N} \mathcal{F}_n$, that is,

$$\mathcal{F}_\infty = \sigma\Big(\bigcup_{n\in N} \mathcal{F}_n\Big).$$

As usual N denotes the set of positive integers and $\bar{N} = N \cup \{+\infty\}$.

We recall that a mapping $\gamma\colon \Omega \to \bar{N}$ is a <u>stopping time</u> if $\{\gamma = n\} \in \mathcal{F}_n$ for each $n \in N$. With the stopping time γ we associate the σ-field

$$\mathcal{F}_\gamma = \{A \in \mathcal{F}_\infty \mid A \cap \{\gamma = n\} \in \mathcal{F}_n, \text{ for each } n \in N\};$$

$\mathcal{F}_\gamma$ is the σ-field of events prior to time γ.

Below we denote by:

T_f the set of all stopping times γ that are <u>finite</u> a.s., that is, such that $P(\{\gamma < +\infty\}) = 1$;

T_b the set of all <u>bounded</u> stopping times.

---

[1]The research of this author is in part supported by the National Science Foundation.

[2]The research of this author was supported in part by Office of Naval Research

Throughout this paper E is a Banach space with norm $\| \ \|$. By an E-valued r.v. we shall always mean a strongly (Bochner) measurable mapping of Ω into E. We denote as usual by $L_E^1 = L_E^1(\Omega,\mathcal{F},P)$ the space of all Bochner integrable E-valued r.v.'s. For $X \in L_E^1$ we write

$$\|X\|_1 = \int_\Omega \|X(\omega)\| dP(\omega) = \int_\Omega \|X\| dP.$$

We say that a set $\mathcal{H} \subset L_E^1$ is L^1-__bounded__ if

$$\sup_{X \in \mathcal{H}} \|X\|_1 < +\infty.$$

If $\mathcal{G} \subset \mathcal{F}$ is a sub-σ-field of $\mathcal{F}$, $E^{\mathcal{G}}$ denotes the __corresponding conditional expectation operator__ in L_E^1.

The notation $(X_n,\mathcal{F}_n)_{n \in N}$ means that the sequence of E-valued r.v.'s $(X_n)_{n \in N}$ is __adapted__ to the sequence of σ-fields $(\mathcal{F}_n)_{n \in N}$ (that is, each X_n is $\mathcal{F}_n$-measurable).

Finally we recall the definition of an "amart" (= asymptotic martingale) (see [6],[5],[9]) and a "uniform amart" (see [2],[3]).

Let $(X_n,\mathcal{F}_n)_{n \in N}$ be such that $X_n \in L_E^1$ for each $n \in N$. Then:

$(X_n,\mathcal{F}_n)_{n \in N}$ is a (strong) __amart__ if[1]

$$\lim_{\tau \in T_b} \int X_\tau dP \quad \text{exists strongly in E;}$$

$(X_n,\mathcal{F}_n)_{n \in N}$ is a __uniform amart__ if

$$\lim_{\substack{(\sigma,\tau) \in T_b \times T_b \\ \sigma \leq \tau}} \|E^{\mathcal{F}_\sigma}[X_\tau] - X_\sigma\|_1 = 0.$$

The object of this paper is to give a characterization of a.s. convergence for an L^1-bounded sequence of r.v.'s (see Theorem 1a, 1b and its Corollary in §3); an earlier, different characterization of a.s. convergence for sequences of real r.v.'s was given in [8].

---

[1]Here and everywhere else in the paper, unless explicitly stated otherwise, the limit is taken as the index of the directed set goes to infinity.

For the purpose of our characterization we must allow unbounded stopping times (the set T_b does not suffice, as the example at the end of §4 shows) and we must introduce the notions of dense set and abundant set of stopping times; these are studied in detail in §2.

§2. Dense sets and abundant sets of stopping times

We note first that the set T_f is a lattice for the natural order relation. In fact, for $\sigma,\tau \in T_f$ we have $\sigma \vee \tau \in T_f$ and $\sigma \wedge \tau \in T_f$.

For a set of stopping times $S \subset T_f$ and an element $\gamma \in T_f$ we introduce the notation:

$$S(\gamma) = \{\sigma \in S \mid \sigma \geq \gamma\},$$

which will come in handy throughout the rest of the paper.

We now formulate the notions of "dense set" and "abundant set" of stopping times:

<u>Definition 1</u>. We say that a set $S \subset T_f$ is <u>dense</u> if: Given any $0 < \varepsilon < 1$ there is $n = n(\varepsilon) \in N$ such that for any $\tau \in T_f$ with $\tau \geq n$, there is $\tau' \in S$ with $P(\{\tau' \neq \tau\}) \leq \varepsilon$.

We next list several properties that a set of stopping times $S \subset T_f$ might satisfy:

a) For each $n \in N$, $S(n) \neq \emptyset$.

b) (Localization) For each finite family $(\tau_j)_{j \in J}$ of stopping times with $\tau_j \in S$ (for $j \in J$) and finite partition of Ω, $(A_j)_{j \in J}$ with $A_j \in \mathcal{F}_{\tau_j}$ (for $j \in J$), if we set $\tau(\omega) = \tau_j(\omega)$ for $\omega \in A_j$, then $\tau \in S$.

c) (Density) The set S is dense.

<u>Definition 2</u>. We say that $S' \subset T_f$ is <u>abundant</u> if S' contains a set S with the properties a), b) and c) above. If S' itself satisfies a), b) and c) we say that S' is <u>abundant in the strict sense</u>.

<u>Remarks</u>. 1) If $S' \subset T_f$ is dense (respectively abundant) and if $S'' \supset S'$, then

S" is dense (respectively abundant); thus both notions defined above are hereditary under upward inclusion. 2) If $S \subset T_f$ satisfies the "localization property" b) above, then clearly S is stable under the operations $\vee$ and $\wedge$, that is, S is a lattice. The converse implication is of course false. 3) In the terminology of [4], if $S \subset T_f$ satisfies conditions a) and b) above, then S is sufficiently rich. Hence to say that S' is abundant means to say that S' contains a set S which is sufficiently rich and dense.

We next collect several basic and useful properties of dense sets and abundant sets of stopping times:

Lemma 1. Let $S \subset T_f$ be dense and stable under the operation $\vee$. Let $\gamma \in T_f$ be such that $S(\gamma) \neq \emptyset$. Then $S(\gamma)$ is dense (and of course stable under the operation $\vee$).

Proof: Let $\theta \in S(\gamma)$ be fixed. Given $0 < \varepsilon < 1$, choose first an integer $p \in N$ such that

$$P(\{\theta \geq p\}) \leq \frac{\varepsilon}{2}.$$

Let $n = n\left(\frac{\varepsilon}{2}\right)$ be the integer corresponding to $\frac{\varepsilon}{2}$ in Definition 1 and let $n' = \max(p,n)$. Let $\tau \in T_f$ with $\tau \geq n'$; there is then $\tau' \in S$ with

$$P(\{\tau' \neq \tau\}) \leq \frac{\varepsilon}{2}.$$

Note that

$$\{\tau' \leq \theta\} \subset \{\theta \geq p\} \cup \{\tau' \neq \tau\}.$$

Let now $\tau'' = \tau' \vee \theta$. Then $\tau'' \in S$ and $\tau'' \geq \gamma$, so that $\tau'' \in S(\gamma)$. Also

$$\{\tau'' \neq \tau\} \subset \{\tau' \neq \tau\} \cup \{\tau' \leq \theta\} \subset \{\tau' \neq \tau\} \cup \{\theta \geq p\}.$$

This implies that $P(\{\tau'' \neq \tau\}) \leq \varepsilon$, completing the proof.

We also have:

Lemma 2. Let $S \subset T_f$ be abundant in the strict sense. Let $\gamma \in T_f$ be such that $S(\gamma) \neq \emptyset$. Then $S(\gamma)$ is also abundant in the strict sense.

Proof: We fix θ in $S(\gamma)$ and we check the properties a), b), c) of Definition 2 for $S(\gamma)$:

a) Note that for each $n \in N$ there is $\sigma \in S$ with $\sigma \geq n$; then $\sigma \vee \theta \in S(\gamma)$ and clearly $\sigma \vee \theta \geq n$.

The localization property b) is obviously satisfied for $S(\gamma)$. The density condition c) now follows from Lemma 1. Hence the lemma is proved.

The following lemma exhibits a whole class of examples of "abundant sets" and is in fact the motivation for introducing the notion of abundant set:

Lemma 3. *Let $(X_n)_{n \in N}$ be an adapted sequence of E-valued r.v.'s. Let $(P(n))_{n \in N}$ be a sequence of adapted properties, i.e. for each $k \in N$ the set*

$$A_k = \{\omega \in \Omega \mid \text{the process } (X_n)_{n \in N} \text{ satisfies } P(k) \text{ at } \omega\}$$

belongs to $\mathcal{F}_k$. Assume that almost surely for $\omega \in \Omega$, the process $(X_n)_{n \in N}$ satisfies $P(k)$ at ω for all k large enough. Let S be the set of all stopping times $\sigma \in T_f$ with the following property: If $k \in N$, the process $(X_n)_{n \in N}$ satisfies $P(k)$ at each $\omega \in \{\sigma = k\}$.

Then the set S is abundant in the strict sense.

Proof: To simplify the terminology, if the process $(X_n)_{n \in N}$ satisfies $P(k)$ at ω, we shall say that $P(k)$ holds at ω.

It is easily verified that the set S satisfies properties a) and b) of Definition 2. It remains to check the density property c):

For each $\tau \in T_f$ define τ' by

$$\tau'(\omega) = \inf\{k \in N \mid k \geq \tau(\omega) \text{ and } P(k) \text{ holds at } \omega\}.$$

It follows from the assumptions that $\tau' \in T_f$. Moreover the relation $\tau \geq n$ implies that the set $\{\tau' = \tau\}$ includes the set

$$B_n = \bigcap_{k \geq n} A_k.$$

For the latter we have by assumption $P(B_n) \nearrow 1$. This clearly finishes the proof.

We recall that a subset $\mathcal{H}$ of L^1_E is *uniformly integrable* if

$$\sup_{f \in \mathcal{H}} \left(\int_{\{\|f\| > a\}} \|f(\omega)\| \, dP(\omega) \right) \to 0 \quad \text{as } a \to +\infty.$$

Alternatively (see [11], p. 17), $\mathcal{H} \subset L^1_E$ is uniformly integrable if and only if the following two conditions are satisfied:

(α) $\sup_{f \in \mathcal{H}} \|f\|_1 < +\infty$, that is, the set $\mathcal{H}$ is L^1-bounded;

and (in the nonatomic case)

(β) for each $\varepsilon > 0$ there is $\delta = \delta(\varepsilon) > 0$ such that the conditions $A \in \mathcal{F}$, $P(A) \leq \delta$ imply

$$\int_A \|f(\omega)\| dP(\omega) \leq \varepsilon \qquad \text{for every } f \in \mathcal{H}.$$

We recall also that if $(f_n)_{n \in N}$ is a sequence of elements of L^1_E that converges almost surely to a r.v. f, then the convergence of f_n to f takes place in the L^1_E-norm if and only if the sequence $(f_n)_{n \in N}$ is uniformly integrable.

The following lemma proved in Neveu (see [12], p. 75) for martingales carries over to our more general setting:

Lemma 4. Let $(X_n)_{n \in N}$ be an adapted sequence of elements of L^1_E; let $\sigma \in \mathbf{T}_f$ and let $(n_k)_{k \in N}$ be a strictly increasing sequence in N. Then the following assertions are equivalent:

(i) The sequence $(X_{\sigma \wedge n_k})_{k \in N}$ is uniformly integrable.

(ii) The r.v. X_σ belongs to L^1_E and the sequence $(X_{\sigma \wedge n_k})_{k \in N}$ converges in L^1_E to X_σ.

(iii) The r.v. X_σ belongs to L^1_E and

$$\lim_k \int_{\{\sigma > n_k\}} \|X_{n_k}(\omega)\| dP(\omega) = 0.$$

Proof: The equivalence (i) $\Leftrightarrow$ (ii) is immediate if one notes that the sequence $(X_{\sigma \wedge n_k})_{k \in N}$ converges almost surely to X_σ.

(i) $\equiv$ (ii) $\Rightarrow$ (iii). It suffices to observe that $P(\{\sigma > n_k\}) \downarrow 0$ as $k \to \infty$ and that

$$\int_{\{\sigma > n_k\}} \|X_{n_k}(\omega)\| dP(\omega) = \int_{\{\sigma > n_k\}} \|X_{\sigma \wedge n_k}(\omega)\| dP(\omega).$$

(iii) ⟹ (ii) follows from the obvious inequality:

$$\int \|X_{\sigma \wedge n_k} - X_\sigma\| dP = \int_{\{\sigma > n_k\}} \|X_{n_k} - X_\sigma\| dP$$

$$\leq \int_{\{\sigma > n_k\}} \|X_{n_k}\| dP + \int_{\{\sigma > n_k\}} \|X_\sigma\| dP.$$

This finishes the proof.

With the notation of the above lemma we have the following immediate corollary.

<u>Corollary</u>. <u>Suppose that the subsequence</u> $(X_{n_k})_{k \in N}$ <u>of</u> $(X_n)_{n \in N}$ <u>is uniformly integrable</u>. <u>Then for each</u> $\sigma \in T_f$ <u>for which</u> $X_\sigma \in L^1_E$ <u>we have</u>

$$X_{\sigma \wedge n_k} \to X_\sigma \quad \underline{\text{in}}\ L^1_E.$$

<u>Definition 3</u>. Let $(X_n)_{n \in N}$ be an adapted sequence of elements of L^1_E. We say that a stopping time $\nu \in T_f$ is <u>regular for</u> $(X_n)_{n \in N}$ if the sequence $(X_{\nu \wedge n})_{n \in N}$ is uniformly integrable.

<u>Remark</u>. The above definition of regular stopping time is standard in martingale theory (see [12], p. 73).

We now give a list of <u>examples of abundant sets of stopping times</u> that arise naturally in the literature. In fact each one of the sets in Examples 1) - 5) below is <u>abundant in the strict sense</u>:

1) $S = T_b$ and $S = T_f$.

2) The class of examples contained in Lemma 3.

3) The set S of all $\sigma \in T_f$ having a first moment, i.e., such that $E(\sigma) < +\infty$.

4) The set S of all $\sigma \in T_f$ satisfying the condition: $P(\{\sigma > n\}) = o(n)$.

5) Let $(X_n)_{n \in N}$ be an adapted sequence of elements of L^1_E and consider the following sets S of stopping times:

5.1) The set S of all stopping times $\nu \in T_f$ such that $X_\nu \in L^1_E$;

5.2) The set S of all stopping times $\nu \in T_f$ that are regular for $(X_n)_{n \in N}$.

We now give some simple examples and counterexamples pertaining to sets of stopping times.

Example. Let $(\Omega,\mathcal{F},P)$ be the Borel space of the unit interval, i.e., $\Omega = [0,1)$, $\mathcal{F}$ = the σ-field of Borel sets, P = Lebesgue measure. For each $n \geq 1$, let $\mathcal{F}_n$ be the σ-field generated by the "dyadic intervals of order n":

$$[0, \frac{1}{2^n}), [\frac{1}{2^n}, \frac{2}{2^n}), \ldots, [\frac{k}{2^n}, \frac{k+1}{2^n}), \ldots, [\frac{2^n - 1}{2^n}, \frac{2^n}{2^n}).$$

Then $(\mathcal{F}_n)_{n \in N}$ is an increasing sequence of σ-fields and $\mathcal{F}_\infty = \sigma(\bigcup_{n \in N} \mathcal{F}_n) = \mathcal{F}$.

(1) Let S be defined as follows: $\sigma \in S$ if j) $\sigma \in T_b$ (that is, σ is bounded); jj) σ is non-constant (that is, σ assumes at least two distinct values); jjj) if $n = \min \sigma$ then $\{\sigma = n\} = [0, \frac{1}{2^n})$. Then S is dense.

(2) Let S' be defined as follows: $\sigma \in S'$ if σ satisfies j) and jj) above and jjj') if $n = \min \sigma$ then the set $\{\sigma = n\}$ coincides with some "dyadic interval of order n," $[\frac{k}{2^n} \frac{k+1}{2^n})$. The set S' contains the set S defined above. Hence S' is dense. However, as is easily seen, the set S' is not stable under the operation $\vee$.

(3) Let S" be defined as follows: $\sigma \in S''$ if σ satisfies j) and jj) above and in addition jjj") if $n = \min \sigma$ then $\{\sigma = n\} \supset [0, \frac{1}{2^n})$. The set S" contains the set S defined above. Hence S" satisfies conditions a) and c) in Definition 2. It is also easily seen that S" is stable under the operation $\vee$, but S" does not satisfy the "localization property" b).

§3. Almost sure convergence in B-spaces

We begin with the following:

Lemma 1. *Let $(Z_n)_{n \in N}$ be an adapted sequence of elements of L^1_E. Assume there is a sequence $(S_n)_{n \in N}$ of sets of stopping times such that each S_n is dense and*

$$\sup_{\sigma \in S_n} \|Z_\sigma\|_1 \to 0 \quad \text{as } n \to \infty.$$

Then, almost surely for $\omega \in \Omega$, $\lim_{n \in N} Z_n(\omega) = 0$ strongly in E.

Proof: For each $n \in N$ define the real r.v. U_n by

$$U_n(\omega) = \|Z_n(\omega)\| \quad \text{for } \omega \in \Omega$$

and let $U^*(\omega) = \limsup_n U_n(\omega)$ for $\omega \in \Omega$. We shall show that $U^*(\omega) = 0$ a.s., which will prove the lemma.

Let $(\varepsilon_n)_{n \in N}$ be a sequence of positive numbers with $\sum_n \varepsilon_n < \infty$. For each $n \in N$ let $p_n \in N$ be such that every $\tau \in T_f$ with $\tau \geq p_n$ can be "approximated to within ε_n" in S_n (density of S_n). Now there is a sequence $(\tau_n)_{n \in N}$ in T_b with $\tau_n \geq p_n$ such that $\lim_n U_{\tau_n}(\omega) = U^*(\omega)$ a.s. (see for instance Theorem 1 in [1]). For each $n \in N$ choose $\sigma_n \in S_n$ with $P(\{\sigma_n \neq \tau_n\}) \leq \varepsilon_n$. Then

$$\lim_n U_{\sigma_n}(\omega) = \lim_n \|Z_{\sigma_n}(\omega)\| = U^*(\omega) \quad \text{a.s.,}$$

and hence, by Fatou's lemma, $\int U^* dP = 0$, proving the lemma.

We now note that if $S \subset T_f$ is a set stable under the operation $\vee$, then S is obviously "<u>directed upward</u>" for the natural order relation.

<u>Corollary</u>. <u>Let</u> $(Z_n)_{n \in N}$ <u>be an adapted sequence of elements of</u> L^1_E. <u>Let</u> $S \subset T_f$ <u>be dense and stable under the operation</u> $\vee$. <u>Assume that</u>

$$\lim_{\sigma \in S} \|Z_\sigma\|_1 = 0.$$

<u>Then, almost surely for</u> $\omega \in \Omega$, $\lim_{n \in N} Z_n(\omega) = 0$ <u>strongly in</u> E.

<u>Proof</u>: For each $n \in N$ choose $\sigma_n \in S$ such that

$$\sigma \in S,\ \sigma \geq \sigma_n \Rightarrow \|Z_\sigma\|_1 \leq \frac{1}{n}.$$

Let $S_n = S(\sigma_n)$ for $n \in N$. Then (use Lemma 1 in §2) the sequence $(S_n)_{n \in N}$ satisfies the assumptions of the previous lemma. This completes the proof.

<u>Remark</u>. Let $S \subset T_f$ be <u>stable</u> under the operation $\vee$. Then the set

$$\Gamma = \{(\sigma,\tau) \in S \times S \mid \sigma \leq \tau\}$$

is "<u>directed upward</u>" for the natural order $(\sigma',\tau') \leq (\sigma'',\tau'')$ if $\sigma' \leq \sigma''$ and $\tau' \leq \tau''$ (in fact, given (σ_1,τ_1) and (σ_2,τ_2) in Γ, if we define $\sigma = \sigma_1 \vee \sigma_2$, $\tau = \tau_1 \vee \tau_2$, then $(\sigma,\tau) \in \Gamma$ and $(\sigma_1,\tau_1) \leq (\sigma,\tau)$, $(\sigma_2,\tau_2) \leq (\sigma,\tau)$). Note now that for <u>any</u> $(\sigma_0,\tau_0) \in \Gamma$ we have:

$$\{(\sigma,\tau) \in \Gamma \mid (\sigma_0,\tau_0) \leq (\sigma,\tau)\} \subset \{(\sigma,\tau) \in \mathbf{S}\times\mathbf{S} \mid \sigma_0 \leq \sigma \leq \tau\}$$

$$= \{(\sigma,\tau) \in \Gamma \mid (\sigma_0,\sigma_0) \leq (\sigma,\tau)\}$$

and

$$\{(\sigma,\tau) \in \mathbf{S}\times\mathbf{S} \mid \tau_0 \leq \sigma \leq \tau\}$$

$$= \{(\sigma,\tau) \in \Gamma \mid (\tau_0,\tau_0) \leq (\sigma,\tau)\} \subset \{(\sigma,\tau) \in \Gamma \mid (\sigma_0,\tau_0) \leq (\sigma,\tau)\}.$$

In particular suppose that $\mathbf{S}$ is as in the previous Remark and $(X_n)_{n\in N}$ is an adapted sequence of E-valued r.v.'s for which the set $\{X_\tau \mid \tau \in \mathbf{S}\}$ is contained in L^1_E. Then the mean convergence relation

$$\lim_{\substack{(\sigma,\tau)\\ \sigma \leq \tau\\ \sigma,\tau\in\mathbf{S}}} \|E^{\mathcal{F}_\sigma}[X_\tau] - X_\sigma\|_1 = 0$$

is equivalent with

$$\lim_{\sigma_0\in\mathbf{S}} \left(\sup_{\substack{(\sigma,\tau)\\ \sigma_0 \leq \sigma \leq \tau\\ \sigma,\tau\in\mathbf{S}}} \|E^{\mathcal{F}_\sigma}[X_\tau] - X_\sigma\|_1 \right) = 0.$$

We may now state the main result of this paper giving a characterization of almost sure convergence. We divide the statement into two parts, Theorem 1a and Theorem 1b.

<u>Theorem 1a</u>. <u>Let</u> E <u>be an arbitrary</u> B-<u>space</u>. <u>Let</u> $(X_n)_{n\in N}$ <u>be an adapted sequence of elements of</u> L^1_E <u>and suppose that</u> $(X_n)_{n\in N}$ <u>is</u> L^1-<u>bounded</u>. <u>Then</u>:

(1) $\lim_{n\in N} X_n(\omega)$ <u>exists in</u> E <u>strongly almost surely</u>

<u>implies</u>

(2) <u>there exists a set</u> $\mathbf{S} \subset \mathbf{T}_f$ <u>which is abundant in the strict sense, such that the set</u> $\{X_\tau \mid \tau \in \mathbf{S}\}$ <u>is</u> L^1-<u>bounded and</u>

$$\lim_{\substack{(\sigma,\tau)\\ \sigma \leq \tau\\ \sigma,\tau\in\mathbf{S}}} \|E^{\mathcal{F}_\sigma}[X_\tau] - X_\sigma\|_1 = 0.$$

<u>Proof</u>: Let $X_\infty(\omega) = \lim_{n\in N} X_n(\omega)$ if the limit exists, and $X_\infty(\omega) = 0$ otherwise. By Fatou's lemma, $X_\infty \in L^1_E$.

Let now $0 < a \leq 1$. We define the "adapted" sequence $(P^{(a)}(k))_{k \in N}$ of measurable properties as follows: We say that the process $(X_n)_{n \in N}$ satisfies $P^{(a)}(k)$ at ω, or simply that $P^{(a)}(k)$ holds at ω, if

$$(P^{(a)}(k)) \qquad \| E^{\mathcal{F}_k}[X_\infty](\omega) - X_k(\omega)\| \leq a.$$

The sequence $(P^{(a)}(k))_{k \in N}$ of measurable properties satisfies the assumptions of Lemma 3 in §2 (this is easily seen, since by the Martingale Convergence Theorem (see [12], p. 104), $\lim_n E^{\mathcal{F}_n}[X_\infty](\omega) = X_\infty(\omega)$ strongly in E, a.s.). Let $S^{(a)}$ be the set of all stopping times $\sigma \in T_f$ with the following property: If $k \in N$, the process $(X_n)_{n \in N}$ satisfies $P^{(a)}(k)$ at each $\omega \in \{\sigma = k\}$. By the lemma quoted above, the set $S^{(a)}$ is abundant in the strict sense.

Define now S to be $S^{(1)}$. It is then clear that S is <u>abundant in the strict sense</u>, that $S \supset S^{(a)}$ for each $0 < a < 1$, and that for each $\tau \in S$ we have

$$\|X_\tau(\omega)\| \leq \|E^{\mathcal{F}_\tau}[X_\infty](\omega)\| + 1, \quad \text{for } \omega \in \Omega;$$

hence

$$\|X_\tau\|_1 \leq \|X_\infty\|_1 + 1$$

and the set $\{X_\tau \mid \tau \in S\}$ is L^1-bounded. We shall now show that

$$\lim_{\substack{(\sigma,\tau) \\ \sigma \leq \tau \\ \sigma,\tau \in S}} \|E^{\mathcal{F}_\sigma}[X_\tau] - X_\sigma\|_1 = 0.$$

Given $0 < \varepsilon < 1$, choose first $a > 0$ such that $a \leq \varepsilon/4$. By Egorov's theorem there are: $A \in \mathcal{F}$ with $P(A^c) \leq a$ and $n_0 \in N$ large enough that:

$$k \geq n_0 \Rightarrow \sup_{\omega \in A} \|E^{\mathcal{F}_k}[X_\infty](\omega) - X_k(\omega)\| \leq a.$$

Define now $\sigma_0 \in T_f$ by:

$$\sigma_0(\omega) = \inf\{k \geq n_0 \mid P^{(a)}(k) \quad \text{holds at } \omega\}.$$

It is clear that $\sigma_0 \in S^{(a)} \subset S$ and that

$$A \subset \{\sigma_0 = n_0\}.$$

Let now $\sigma,\tau \in \mathbf{S}$ with $\tau \geq \sigma \geq \sigma_0$; we have

$$\omega \in A \Rightarrow \begin{cases} \| E^{\mathcal{F}_\sigma}[X_\infty](\omega) - X_\sigma(\omega)\| \leq a \\ \| E^{\mathcal{F}_\tau}[X_\infty](\omega) - X_\tau(\omega)\| \leq a \end{cases}$$

and

$$\omega \in A^c \Rightarrow \begin{cases} \| E^{\mathcal{F}_\sigma}[X_\infty](\omega) - X_\sigma(\omega)\| \leq 1 \\ \| E^{\mathcal{F}_\tau}[X_\infty](\omega) - X_\tau(\omega)\| \leq 1 \end{cases}$$

and $P(A^c) \leq a$. From these relations we deduce

$$\| E^{\mathcal{F}_\sigma}[X_\infty] - X_\sigma\|_1 \leq a + a \leq \frac{\varepsilon}{2}$$

$$\| E^{\mathcal{F}_\tau}[X_\infty] - X_\tau\|_1 \leq a + a \leq \frac{\varepsilon}{2}.$$

Applying the projection operator $E^{\mathcal{F}_\sigma}$ to the last inequality we get (since $E^{\mathcal{F}_\sigma}(E^{\mathcal{F}_\tau}) = E^{\mathcal{F}_\sigma}$)

$$\| E^{\mathcal{F}_\sigma}[X_\infty] - E^{\mathcal{F}_\sigma}[X_\tau]\|_1 \leq \frac{\varepsilon}{2}$$

and finally, by the triangle inequality

$$\| E^{\mathcal{F}_\sigma}[X_\tau] - X_\sigma\|_1 \leq \varepsilon.$$

Since this holds for every pair (σ,τ) with $\sigma,\tau \in \mathbf{S}$ and $\tau \geq \sigma \geq \sigma_0$, the implication (1) $\Rightarrow$ (2) is proved. Hence Theorem 1a is proved.

If the B-space E <u>has the Radon-Nikodym property,</u> we say for short that E has RNP (see [7] or [12], p. 112).

<u>Theorem 1b</u>. <u>Assume that the</u> B-<u>space</u> E <u>has</u> RNP. <u>Let</u> $(X_n)_{n \in N}$ <u>be an adapted sequence of</u> E-<u>valued</u> r.v.'s. <u>Then</u>:

(3) <u>There exists a decreasing sequence</u> $(\mathbf{S}_n)_{n \in N}$ <u>of sets of stopping times such that</u>: $\mathbf{S}_n \subset \mathbf{T}_f(n)$, $\mathbf{S}_n$ <u>is dense, the set</u> $\{X_\tau \mid \tau \in \mathbf{S}_n\}$ <u>is</u> L^1-<u>bounded and</u>

$$\sup_{\substack{(\sigma,\tau)\\ \sigma,\tau\in \mathbf{S}_n}} \| E^{\mathcal{F}_\sigma}[X_\tau - X_\sigma]\cdot 1_{\{\sigma\le\tau\}}\|_1 \to 0 \quad \underline{\text{as}}\ n\to\infty\}$$

<u>implies statement</u> (1) <u>in Theorem 1a</u>.

<u>Proof</u>: Let the sequence $(\mathbf{S}_n)_{n\in N}$ be as in (3) and for each $n \in N$ let

$$\varepsilon_n = \sup_{\substack{(\sigma,\tau)\\ \sigma,\tau\in \mathbf{S}_n}} \| E^{\mathcal{F}_\sigma}[X_\tau - X_\sigma]\cdot 1_{\{\sigma\le\tau\}}\|_1.$$

To prove the implication we show that we have a decomposition $X_n = Y_n + Z_n$, for $n \in N$, where $(Y_n)_{n\in N}$ is an L^1-bounded martingale and $(Z_n)_{n\in N}$ satisfies the assumptions of Lemma 1 above.

<u>Fix</u> $n \in N$. For $p \ge n$ and arbitrary $\sigma,\tau \in \mathbf{S}_p$ we have (since $E^{\mathcal{F}_n}(E^{\mathcal{F}_\sigma}) = E^{\mathcal{F}_n}$, $E^{\mathcal{F}_n}(E^{\mathcal{F}_\tau}) = E^{\mathcal{F}_n}$ and $\{\sigma\le\tau\}\in\mathcal{F}_\sigma$, $\{\tau\le\sigma\}\in\mathcal{F}_\tau$)

$$\begin{aligned} &\| E^{\mathcal{F}_n}[X_\tau] - E^{\mathcal{F}_n}[X_\sigma]\|_1 \\ &\le \| E^{\mathcal{F}_n}[(X_\tau - X_\sigma)\cdot 1_{\{\sigma\le\tau\}}]\|_1 + \| E^{\mathcal{F}_n}[(X_\tau - X_\sigma)\cdot 1_{\{\tau\le\sigma\}}]\|_1 \\ &\le \| E^{\mathcal{F}_\sigma}[X_\tau - X_\sigma]\cdot 1_{\{\sigma\le\tau\}}\|_1 + \| E^{\mathcal{F}_\tau}[X_\sigma - X_\tau]\cdot 1_{\{\tau\le\sigma\}}\|_1 \le 2\varepsilon_p. \end{aligned}$$

Thus we have

$$(1^\circ)\qquad \left.\begin{matrix} p\ge n\\ \sigma,\tau\in\mathbf{S}_p\end{matrix}\right\} \Rightarrow \| E^{\mathcal{F}_n}[X_\tau] - E^{\mathcal{F}_n}[X_\sigma]\|_1 \le 2\varepsilon_p.$$

For each $p \ge n$, let

$$G_p(n) = \{E^{\mathcal{F}_n}[X_\tau] \mid \tau\in\mathbf{S}_p\}$$

and

$$\overline{G_p(n)} = \text{closure of } G_p(n) \text{ in } L^1_E.$$

By (1°) above $(\overline{G_p(n)})_{p\ge n}$ is a decreasing sequence of closed sets in L^1_E with diameter tending to zero. Define Y_n to be (the unique element)

$$Y_n \in \bigcap_{p \geq n} \overline{G_p(n)}.$$

Because of (1°) we obtain

$$\text{(2°)} \qquad \left.\begin{matrix} p \geq n \\ \sigma \in S_p \end{matrix}\right\} \Rightarrow \|E^{\mathcal{F}_n}[X_\sigma] - Y_n\|_1 \leq 2\varepsilon_p.$$

It is now easy to check that $(Y_n)_{n \in N}$ is a martingale. In fact, let $m > n$. For each $p \geq m$ choose $\sigma_p \in S_p$. By (2°) we have

$$E^{\mathcal{F}_m}[X_{\sigma_p}] \xrightarrow[p \to \infty]{} Y_m \quad \text{in } L^1_E$$

and applying the projection operator $E^{\mathcal{F}_n}$,

$$E^{\mathcal{F}_n}[X_{\sigma_p}] \xrightarrow[p \to \infty]{} E^{\mathcal{F}_n}[Y_m] \quad \text{in } L^1_E.$$

On the other hand also by (2°) above

$$E^{\mathcal{F}_n}[X_{\sigma_p}] \xrightarrow[p \to \infty]{} Y_n \quad \text{in } L^1_E.$$

This shows that $Y_n = E^{\mathcal{F}_n}[Y_m]$, i.e. $(Y_n)_{n \in N}$ is a martingale. The L^1-boundedness of $(Y_n)_{n \in N}$ is an immediate consequence of (2°).

We now set

$$Z_n = X_n - Y_n, \quad \text{for } n \in N,$$

and we show that

$$\text{(3°)} \qquad \left.\begin{matrix} \sigma \in S_n \\ p \geq n \end{matrix}\right\} \Rightarrow \int_{\{\sigma \leq p\}} \|Z_\sigma\| dP \leq 2\varepsilon_n.$$

<u>Fix</u> $\sigma \in S_n$ <u>and</u> $p \geq n$. Choose $q > p$ large enough that $2p\varepsilon_q \leq \varepsilon_n$. <u>Choose</u> $\tau \in S_q$. By (2°) above we have for each $j \leq p$,

$$\|E^{\mathcal{F}_j}[X_\tau] - Y_j\|_1 \leq 2\varepsilon_q$$

which implies

$$(4°)\qquad \|1_{\{\sigma \leq p\}}(Y_\sigma - E^{\mathcal{F}_\sigma}[X_\tau])\|_1 \leq \sum_{j=1}^{p} \int_{\{\sigma = j\}} \|Y_j - E^{\mathcal{F}_j}[X_\tau]\| dP$$

$$\leq p(2\varepsilon_q) \leq \varepsilon_n.$$

On the other hand $\tau \geq p$ and both σ and τ belong to S_n; we deduce

$$(5°)\qquad \|1_{\{\sigma \leq p\}}(X_\sigma - E^{\mathcal{F}_\sigma}[X_\tau])\|_1 = \|E^{\mathcal{F}_\sigma}[X_\tau - X_\sigma]\cdot 1_{\{\sigma \leq p\}}\|_1$$

$$\leq \|E^{\mathcal{F}_\sigma}[X_\tau - X_\sigma]\cdot 1_{\{\sigma \leq \tau\}}\|_1 \leq \varepsilon_n.$$

Inequality (3°) now follows from (4°) and (5°).

If we let $p \to \infty$ in inequality (3°) and we note that $\{\sigma \leq p\} \uparrow \{\sigma < +\infty\}$, we get

$$(6°)\qquad \sigma \in S_n \Longrightarrow \int_\Omega \|Z_\sigma\| dP \leq 2\varepsilon_n.$$

Thus $(Z_n)_{n\in N}$ and $(S_n)_{n\in N}$ satisfy the assumptions of Lemma 1.

The proof of the implication (3) $\Longrightarrow$ (1) is now concluded by applying Lemma 1 to $(Z_n)_{n\in N}$ and the Martingale Convergence Theorem to $(Y_n)_{n\in N}$ (this is possible since E has RNP; see [7] or [12], p. 112). This proves Theorem 1b.

Corollary. *In particular for B-spaces having RNP the assertions (1), (2), (3) are equivalent.*

Proof: Since (1) $\Longrightarrow$ (2) by Theorem 1a and (3) $\Longrightarrow$ (1) by Theorem 1b, it suffices to note that the implication (2) => (3) is always valid. In fact, let S be as in (2). Define by induction an increasing sequence of stopping times $(\sigma_n)_{n\in N}$ with $\sigma_n \in S$, $\sigma_n \geq n$ such that

$$\left.\begin{array}{r}\sigma,\tau \in S\\ \sigma_n \leq \sigma \leq \tau\end{array}\right\} \Longrightarrow \|E^{\mathcal{F}_\sigma}[X_\tau] - X_\sigma\|_1 \leq \frac{1}{n}$$

and let $S_n = S(\sigma_n)$. Then the sequence $(S_n)_{n\in N}$ is decreasing, S_n is abundant in the strict sense (see Lemma 2 in §2) and the set $\{X_\tau \mid \tau \in S_n\}$ is L^1-bounded.

Let now σ,τ be arbitrary elements in S_n. Then $\sigma \vee \tau \in S_n$, the set $\{\sigma \leq \tau\}$ belongs to $\mathcal{F}_\sigma$ and on this set $\tau = \sigma \vee \tau$ so that

$$\|E^{\mathcal{F}_\sigma}[X_\tau - X_\sigma]\cdot 1_{\{\sigma \leq \tau\}}\|_1 \leq \|E^{\mathcal{F}_\sigma}[X_{\sigma \vee \tau}] - X_\sigma\|_1 \leq \frac{1}{n}.$$

This completes the proof of the implication (2) => (3) and hence of the Corollary.

Remarks. 1) One may formulate several other assertions that are "sandwiched in between" the assertions (2) and (3) of Theorem 1. For instance:

(2') There exists a set $S \subset T_f$ such that $S(n) \neq \emptyset$ for each $n \in N$, S is stable under the operation $\vee$ and dense, the set $\{X_\tau \mid \tau \in S\}$ is L^1-bounded and

$$\lim_{\substack{(\sigma,\tau)\\ \sigma \leq \tau\\ \sigma,\tau \in S}} \|E^{\mathcal{F}_\sigma}[X_\tau] - X_\sigma\|_1 = 0.$$

(2") There exists a decreasing sequence $(S_n)_{n\in N}$ of sets of stopping times such that: $S_n \subset T_f(n)$, S_n is-stable under the operation $\vee$ and dense, the set $\{X_\tau \mid \tau \in S_n\}$ is L^1-bounded and

$$\sup_{\substack{(\sigma,\tau)\\ \sigma \leq \tau\\ \sigma,\tau \in S_n}} \|E^{\mathcal{F}_\sigma}[X_\tau] - X_\sigma\|_1 \to 0 \quad \text{as } n \to \infty.$$

Theorem 1b may now be reformulated as follows: In Banach spaces having RNP every one of the implications (2) => (1), (2') => (1), (2") => (1), (3) => (1) holds. This of course generalizes the "uniform amart" convergence theorem: take $S = T_b$ in (2) (see [2] and [3], p. 63-65). 2) As a matter of fact the validity of either one of the implications (2) => (1), (2') => (1), (2") => (1), (3) => (1) characterizes the class of Banach spaces having RNP. 3) The Corollary of Theorem 1 may also be reformulated as follows: For Banach spaces having RNP the assertions (1), (2), (2'), (2"), (3) are all equivalent.

We need one more lemma (see also [4], Lemma 1):

Lemma 2. *Let $(X_n)_{n\in N}$ be an adapted sequence of elements of L^1_E. Let $S \subset T_f$ satisfy the "localization property" b) and be such that the set $\{X_\tau \mid \tau \in S\}$ is contained in L^1_E. Then we have:*

I) $$\sup_{\substack{(\sigma,\tau)\\ \sigma,\tau\in S}} \left\|\int (X_\tau - X_\sigma)dP\right\| \le \sup_{\substack{(\sigma,\tau)\\ \sigma\le\tau\\ \sigma,\tau\in S}} \|E^{\mathcal{F}_\sigma}[X_\tau] - X_\sigma\|_1 .$$

In the particular case when $E = R$ we have equality:

II) $$\sup_{\substack{(\sigma,\tau)\\ \sigma,\tau\in S}} \int (X_\tau - X_\sigma)dP = \sup_{\substack{(\sigma,\tau)\\ \sigma\le\tau\\ \sigma,\tau\in S}} \|E^{\mathcal{F}_\sigma}[X_\tau] - X_\sigma\|_1 .$$

Proof: To prove the inequality I) let $\sigma,\tau \in S$ and let $A = \{\sigma \le \tau\}$. Then $\sigma \vee \tau \in S$ and $\sigma\wedge\tau \in S$ (see Remark 2), following Definition 2 in §2). Also $A \in \mathcal{F}_{\sigma\wedge\tau}$ (since $A = \{\sigma\wedge\tau = \sigma\}$). Therefore

$$\int (X_\tau - X_\sigma)dP = \int_A (X_{\sigma\vee\tau} - X_{\sigma\wedge\tau})dP + \int_{A^c} (X_{\sigma\wedge\tau} - X_{\sigma\vee\tau})dP$$

$$= \int_A (E^{\mathcal{F}_{\sigma\wedge\tau}}[X_{\sigma\vee\tau}] - X_{\sigma\wedge\tau})dP + \int_{A^c} (X_{\sigma\wedge\tau} - E^{\mathcal{F}_{\sigma\wedge\tau}}[X_{\sigma\vee\tau}])dP,$$

whence

$$\left\|\int (X_\tau - X_\sigma)dP\right\| \le \|E^{\mathcal{F}_{\sigma\wedge\tau}}[X_{\sigma\vee\tau}] - X_{\sigma\wedge\tau}\|_1 .$$

This proves the first part of the lemma.

Assume now that $E = R$. In this case it is clear that

$$\sup_{\substack{(\sigma,\tau)\\ \sigma,\tau\in S}} \int (X_\tau - X_\sigma)dP = \sup_{\substack{(\sigma,\tau)\\ \sigma,\tau\in S}} \left|\int (X_\tau - X_\sigma)dP\right| .$$

Hence to prove II) we need only show that

$$\sup_{\substack{(\sigma,\tau)\\ \sigma\le\tau\\ \sigma,\tau\in S}} \|E^{\mathcal{F}_\sigma}[X_\tau] - X_\sigma\|_1 \le \sup_{\substack{(\sigma,\tau)\\ \sigma,\tau\in S}} \int (X_\tau - X_\sigma)dP.$$

To see this, let $\sigma,\tau \in S$, $\sigma \le \tau$ and let $B = \{X_\sigma \le E^{\mathcal{F}_\sigma}[X_\tau]\}$. Clearly $B, B^c \in \mathcal{F}_\sigma \subset \mathcal{F}_\tau$.

Hence if we define σ', τ' by

$$\sigma'(\omega) = \begin{cases} \sigma(\omega) & \text{for } \omega \in B \\ \tau(\omega) & \text{for } \omega \in B^c \end{cases} \qquad \tau'(\omega) = \begin{cases} \tau(\omega) & \text{for } \omega \in B \\ \sigma(\omega) & \text{for } \omega \in B^c \end{cases}$$

then $\sigma',\tau' \in \mathbf{S}$ and we have

$$\|E^{\mathcal{F}_\sigma}[X_\tau] - X_\sigma\|_1 = \int_B (E^{\mathcal{F}_\sigma}[X_\tau] - X_\sigma)dP + \int_{B^c} (X_\sigma - E^{\mathcal{F}_\sigma}[X_\tau])dP$$

$$= \int_B (X_\tau - X_\sigma)dP + \int_{B^c} (X_\sigma - X_\tau)dP = \int (X_{\tau'} - X_{\sigma'})dP;$$

this proves the above assertion and thus completes the proof of Lemma 2.

§4. The real-valued case

We begin with the following lemma which is an easy consequence of Lemma 2 in §3:

Lemma 1. Let $(X_n)_{n \in N}$ be an adapted sequence of elements of L^1_R. Let $\mathbf{S} \subset \mathbf{T}_f$ satisfy the "localization property" b) and be such that the set $\{X_\tau \mid \tau \in \mathbf{S}\}$ is contained in L^1_R. Then we have:

$$\lim_{\substack{(\sigma,\tau) \\ \sigma \leq \tau \\ \sigma,\tau \in \mathbf{S}}} \|E^{\mathcal{F}_\sigma}[X_\tau] - X_\sigma\|_1 = 0$$

if and only if

$$\lim_{\sigma \in \mathbf{S}} \int X_\sigma dP \quad \text{exists in } R.$$

Proof: It suffices to note that, by Lemma 2 in §3, for $0 < \varepsilon < 1$ and $\sigma_0 \in \mathbf{S}$ (the set $\mathbf{S}(\sigma_0)$ clearly satisfies the "localization property" b)), the relation

$$\sup_{\substack{(\sigma,\tau) \\ \sigma \leq \tau \\ \sigma,\tau \in \mathbf{S}(\sigma_0)}} \|E^{\mathcal{F}_\sigma}[X_\tau] - X_\sigma\|_1 \leq \varepsilon$$

is equivalent to

$$\sup_{\substack{(\sigma,\tau) \\ \sigma,\tau \in \mathbf{S}(\sigma_0)}} \left|\int (X_\tau - X_\sigma)dP\right| \leq \varepsilon$$

(the latter is just the "Cauchy condition" for the net of real numbers $(\int X_\sigma dP)_{\sigma \in S}$). Thus the lemma is proved.

We may now state the following variant of Theorem 1 in §3, in the real-valued case:

Theorem 2. *Let $(X_n)_{n \in N}$ be an adapted sequence of elements of L^1_R and suppose that $(X_n)_{n \in N}$ is L^1-bounded. Then the following are equivalent assertions:*

(1) *$\lim_{n \in N} X_n(\omega)$ exists in R almost surely.*

(2) *There exists a set $S \subset T_f$ which is abundant in the strict sense, such that the set $\{X_\tau \mid \tau \in S\}$ is L^1-bounded and*

$$\lim_{\sigma \in S} \int X_\sigma dP \quad \text{exists in } R.$$

This is an immediate consequence of Theorem 1 in §3 and of the previous lemma.

Remark. Theorem 2 contains of course the (real-valued) "amart" convergence theorem: take $S = T_b$ in (2) (see for instance [1] or [9]).

We conclude with an example showing that the set T_b does not suffice for the purposes of Theorem 1 (or, for that matter, Theorem 2) and that we must allow sets of unbounded stopping times.

An example. Let $(\Omega,\mathcal{F},P)$ be the Borel space of the unit interval, i.e. $\Omega = [0,1)$, $\mathcal{F}$ = the σ-field of Borel sets, P = Lebesgue measure. For each $n \geq 1$ let

$$I_n = [\frac{1}{n+1}, \frac{1}{n}),$$

let

$$X_n = (n+1)^2 \cdot 1_{I_n}$$

and

$$\mathcal{F}_n = \sigma(X_1,\ldots,X_n).$$

It is clear that $(X_n)_{n \in N}$ is L^1-bounded and that $\lim_{n \in N} X_n(\omega) = 0$ a.s. Note also that $\mathcal{F}_n$ is spanned by $I_1, I_2,\ldots,I_n$ and hence the set $[0, \frac{1}{n+1})$ is an atom of $\mathcal{F}_n$.

Take now $\sigma \in T_b$ and let $\sigma(0) = p$. Then the set $\{\sigma = p\}$ contains 0 and belongs to $\mathcal{F}_p$ and therefore

$$A = \left[0, \frac{1}{p+1}\right) \subset \{\sigma = p\}.$$

Now A is an atom of $\mathcal{F}_p$ and hence also an atom of $\mathcal{F}_\sigma$ (for a set $B \subset \{\sigma = p\}$ we have $B \in \mathcal{F}_\sigma \iff B \in \mathcal{F}_p$).

Let now τ be <u>any</u> element in T_b satisfying $\tau > \max \sigma = n \geq p$. Clearly,

$$\text{Support of } X_\tau \subset \left[0, \frac{1}{n+1}\right) \subset A,$$

so that

$$E^{\mathcal{F}_\sigma}[X_\tau](\omega) = \frac{\int X_\tau \, dP}{P(A)} \qquad \text{for } \omega \in A.$$

On the other hand

$$X_\sigma(\omega) = 0 \qquad \text{for } \omega \in A.$$

We obtain

$$\| E^{\mathcal{F}_\sigma}[X_\tau] - X_\sigma \|_1 \geq \int_A | E^{\mathcal{F}_\sigma}[X_\tau] - X_\sigma | \, dP = \int_A E^{\mathcal{F}_\sigma}[X_\tau] \, dP = \int X_\tau \, dP > 1.$$

Thus <u>no</u> set $S \subset T_b$ can satisfy assertion (2) in Theorem 1.

References

[1] Austin, D. G., Edgar, G. A., and Ionescu Tulsea, A., "Pointwise convergence in terms of expectations," Zeit. Wahrs. verw. Gebiete, 30, pp. 17-26 (1974).

[2] Bellow, A., "Uniform amarts: A class of asymptotic martingales for which strong almost sure convergence obtains," Zeit. Wahrs. verw. Gebiete, 41, pp. 177-191 (1978).

[3] Bellow, A., "Some aspects of the theory of vector-valued amarts," Proc. Dublin Conference 1977, Vector space measures and applications I, Lecture Notes in Math. No. 644, pp. 57-67, Springer-Verlag (1978).

[4] Bellow, A., "Sufficiently rich sets of stopping times, measurable cluster points and submartingales," Séminaire sur la géométrie des espaces de Banach, École Polytechnique 1977-1978, pp. A.1-A.11.

[5] Brunel, A., and Sucheston, L., "Sur les amarts à valeurs vectorielles," C. R. Acad. Sci. Paris, 283, Série A, pp. 1037-1039 (1976).

[6] Chacon, R. V. and Sucheston, L., "On convergence of vector-valued asymptotic martingales," Zeit. Wahrs. verw. Gebiete, 33, pp. 55-59 (1975).

[7] Chatterji, S. D., "Martingale convergence and the Radon-Nikodym theorem in Banach spaces," Math. Scandinavica, 22, pp. 21-41 (1968).

[8] Dvoretzky, A., "On stopping time directed convergence," Bull. Amer. Math. Soc., 82, No. 2, pp. 347-349 (1976).

[9] Edgar, G. A., and Sucheston, L., "Amarts: A class of asymptotic martingales (Discrete parameter)," J. Multivariate Anal., 6, pp. 193-221 (1976).

[10] Edgar, G. A., and Sucheston, L., "Martingales in the limit and amarts," Proc. Amer. Math. Soc., 67, pp. 315-320 (1977).

[11] Meyer, P. A., Probability and potentials, Blaisdell, Waltham, Mass. 1966.

[12] Neveu, J., Martingales à temps discret, Masson, Paris, 1972.

BANACH SPACE VALUED GAUSSIAN PROCESSES

by

René CARMONA

Département de Mathématiques
Université de Saint-Etienne
23 rue du Docteur P.Michelon
42100 SAINT ETIENNE
FRANCE

I. INTRODUCTION

Let B be a real separable Banach space, and let T be some set. A B-valued Gaussian process on T is a collection $X=\{X(t);t\in T\}$ of B-valued random variables defined on some complete probability space $(\Omega,\mathcal{A},P)$ such that for each finite subset $\{t_1,..,t_n\}$ of T and each finite subset $\{x'_1,..,x'_n\}$ of the dual B' of B, ([1]) with the same cardinality, the real valued random variable $\langle x'_1,X(t_1)\rangle+..+\langle x'_n,X(t_n)\rangle$ is Gaussian ([2]). The covariance of the process, say Γ, is defined by:

$$\Gamma_{(s,t)}(x',y') = E\{\langle x',X(s)\rangle\langle y',X(t)\rangle\} \qquad s,t\in T,\ x',y'\in B' \quad (1,1)$$

As in the real valued case the distribution of the process is uniquely determined by its covariance. So, it is natural to ask for conditions on the covariance that insure the existence of the process and regularity properties of the sample paths. For example, if such a process exist and

a) if d is a metric on T, what kind of assumptions on Γ will imply that P-almost all the paths are continuous functions on the metric space (T,d)? Or

b) if $\mathcal{T}$ is a σ-field of subsets of T and if λ is a σ-finite measure on $(T,\mathcal{T})$,

---

([1]) We use the symbol $\langle\ ,\ \rangle$ to denote the duality between B and B'.

([2]) By Gaussian we mean in fact mean zero Gaussian

what kind of assumptions on Γ will imply that P-almost all the paths belong to some Lebesgue space (or some Orlicz space) constructed on $(\Omega,\mathcal{Q},P)$?

In the real valued case, these questions have been intensively investigated and satisfactory results are known. Surprisingly, the sufficient conditions which are known in the general case are most of the time of no use because they are too complicated and too difficult to check. The reason why the situation is so bad has much to do with the highly complicated structure of the covariance . For example let us point out that Γ has to satisfy: i) $\Gamma_{(s,s)}$ and $\Gamma_{(t,t)}$ are covariances of Gaussian measures on B ii) there is a Gaussian measure on $B \times B$ which is the joint distribution of two B-valued Gaussian random variables the covariances of which are $\Gamma_{(s,s)}$ and $\Gamma_{(t,t)}$, and which are correlated via the bilinear form $\Gamma_{(s,t)}$. And the necessity of conditions i) and ii) contributes to make the statements more involved than usual.

Of particular interest is the simplest case when Γ is the product of the covariance of a real valued process and the covariance of a Gaussian measure on B. Indeed this is essentially the so-called problem of tensor products of abstract Wiener spaces. Section II is devoted to a brief survey of the present state of affairs. In section III we prove a theorem for more general covariances. This result is a typical answer to problem b) above, and it gives a new proof of a recently obtained result on tensor products of abstract Wiener spaces.

Notation: we use the symbol d_Y to denote the pseudo-distance

$$d_Y(s,t) = E\left\{|Y(s)-Y(t)|^2\right\}^{1/2} \qquad s,t\in T$$

whenever $Y=\{Y(t);t\in T\}$ is a real valued Gaussian process.

II. TENSOR PRODUCTS OF ABSTRACT WIENER SPACES

The concept of abstract Wiener space was introduced in th sixties by L.Gross.

Definition([6])

An abstract Wiener space is a triplet (i,H,B) where H is a real separable Hilbert space, B is a real separable Banach space and i is a one-to-one continuous linear map from H into B with dense range which maps the canonical Gaussian cylindrical measure of H, say γ_H, into a σ-additive Borel probability measure on B.

Let (i,H,B) and (j,K,C) be two abstract Wiener spaces. $i\otimes j$ is a linear map from the algebraic tensor product $H\otimes K$ into the algebraic tensor product $B\otimes C$. Moreover it is well known that $H\otimes K$ can be eqipped with a cross norm so that its completion, say $H\hat{\otimes}_2 K$, is a real separable Hilbert space. An interesting stability problem is then the following:

$(\mathcal{P})$ "is $(i\otimes j, H\hat{\otimes}_2 K, B\hat{\otimes}_\alpha C)$ an abstract Wiener space whenever α is a reasonnable norm [3] on $B\otimes C$? "

where $B\hat{\otimes}_\alpha C$ denotes the completion of $B\otimes C$ with respect to the norm α. If there is no extra assumption on α there is no a priori reason why the linear map $i\otimes j$ should possess a continuous one-to-one extension from $H\hat{\otimes}_2 K$ into $B\hat{\otimes}_\alpha C$ with dense range. So the first problem is to prove that:

(i)- *the setting of the problem is meaningful.*

This is indeed the case because: a) $i\otimes j$ extends to a continuous linear map from $H\hat{\otimes}_2 K$ into $B\hat{\otimes}_\alpha C$ because $i(\gamma_H)$ and $j(\gamma_K)$ are σ-additive (see [3.Proposition 3.1]), b) to each element of $B\otimes C$ we can associate a finite rank operator from B' into C, and this association extends to a continuous linear map, say θ, from $B\hat{\otimes}_\alpha C$ into L(B',C) the Banach space of bounded linear operators from B' into C.Now, each $h\in H\hat{\otimes}_2 K$ can be viewed as a Hilbert-Schmidt operator from H into K, and once we identify H and its dual H' we have:

$$[\theta\circ(i\otimes j)](h) = j\circ h\circ i'$$

which shows that $\theta\circ(i\otimes j)$ is one-to-one, and thus so is $i\otimes j$ (we must keep worrying about the fact that θ is not one-to-one in general). The next problem to solve is:

(ii)- *is the answer to $(\mathcal{P})$ yes in full generality,?*

There is a counter-example (the proof of which is due to Gilles Pisier) which shows that if $\alpha=\pi$ the strongest cross norm, the answer can be NO even if B and C are Hilbert spaces (see [3.Section III]). Nevertheless

(iii)- *the answer to $(\mathcal{P})$ is YES for $\alpha=\varepsilon$ the weakest cross norm.*

Chronologically this was the first result on the problem we consider here. It is due to Simone Chevet [4].Surprisingly it appeared a long time after the introduction of the concept of abstract Wiener space. Since there is a bunch of reasonnable norms which are weaker than the projective one π, and stronger than the inductive one ε, an interesting problem is to classify the reasonnable norms and the Banach spaces according to the answer to problem $(\mathcal{P})$. A first contribution to this program is:

(iv)- *the answer to $(\mathcal{P})$ is YES whenever $C = L^p(T,\mathcal{T},\lambda)$ for some $1\leq p<\infty$ and some σ-finite measure space $(T,\mathcal{T},\lambda)$ and $\alpha=\Delta_p$ the norm induced on*

---

[3] a norm α on $B\otimes C$ is called reasonnable if

$$\alpha(x\otimes y) = \|x\|\,\|y\|$$

$$|\langle x'\otimes y', u\rangle| \leq \alpha(u)\,\|x'\|\,\|y'\|$$

whenever $x\in B$, $y\in C$, $x'\in B'$, $y'\in C'$ and $u\in B\otimes C$.

$B\otimes L^p(T,\mathcal{T},\lambda)$ by the norm of $L^p(T,\mathcal{T},\lambda;B)$.

This result is proved in [3] where some corollaries are given. In the next section we give a different proof: in the same way S.Chevet's result reduces to proving that a Banach space valued Gaussian process exists and has almost surely continuous paths, proving (iv) reduces to proving that a Banach space valued Gaussian process exists and to checking that the paths have norms whose p^{th}-powers are almost surely integrable on $(T,\mathcal{T},\lambda)$.

III. INTEGRABILITY OF PATHS OF BANACH SPACE VALUED GAUSSIAN PROCESSES

The first result of this section is a straightforward generalization of the corresponding result in the real valued case [7]. From now on, B is a fixed real separable Banach space and $(T,\mathcal{T},\lambda)$ is a fixed σ-finite countabily generated measure space and $p\in[1,\infty[$.

Proposition

Let $X = \{X(t);t\in T\}$ be a measurable (4) B-valued Gaussian process and let us set $\rho(t) = E\{\|X(t)\|^2\}$ for $t\in T$. Then the paths of X have norms whose p^{th}-powers are almost surely integrable if and only if $\int_T \rho(t)^{p/2} d\lambda(t)<\infty$

Proof:

The proof of the real valued case applies to the present situation once one remarks that: a) integrability properties of Gaussian random variables remain true in the Banach space valued case [5.Section1.3] b) the following estimate due to J.Hoffmann Jørgensen is true (see for example [1.p.112-113]):

$$c_p\, E\{\|Z\|^2\}^{1/2} \leq E\{\|Z\|^p\}^{1/p} \leq C_p\, E\{\|Z\|^2\}^{1/2}$$

where c_p and C_p are universal constants and Z any Banach space valued Gaussian random variable ■

Now we are in a position to prove the main result.

Theorem

Let $\Gamma=\{\Gamma_{(s,t)};\ (s,t)\in T\times T\}$ be a collection of bilinear forms on B' so that the mapping:

$$(T\times B')\times(T\times B') \ni ((s,x'),(t,y')) \longmapsto \Gamma_{(s,t)}(x',y') \qquad (3.1)$$

(4) i.e. the mapping $T\times\Omega \ni (t,\omega) \longmapsto X(t,\omega)\in B$ is $\mathcal{T}\times\mathcal{A}$-measurable.

is a symmetric kernel of positive type which is $\mathcal{T}\times\mathcal{B}_{\sigma(B')}$ *measurable* (5) *for each fixed* $(s,x')\in T\times B'$. *Furthermore let us assume the existence of a* $\varphi\in L^p(T,\mathcal{T},\lambda)$ *and of a continuous sample paths real valued Gaussian process* $Z=\{Z(u);u\in U\}$ *where* U *is a closed ball in* B' *centered at the origin and equipped with the weak*topology, so that*

$$\forall t\in T,\ \forall (u,u')\in U\times U,\quad \Gamma_{(t,t)}(u-u',u-u')^{1/2}\le \varphi(t)\ E\{|Z(u)-Z(u')|^2\}^{1/2} \tag{3.2}$$

Then, there exists a measurable B-valued Gaussian process $X=\{X(t);t\in T\}$ *the sample paths of which have norms whose* p^{th}*-powers are almost surely integrable, and which satisfies* (1.1).

Remark:

It is clear that it is sufficient to check the positivity of (3.1) for x' *and* y' *with a priori given norms.*

Proof:

From classical results on orthogonal expansions of Gaussian processes, we can find a sequence $\{\xi_n;n\ge 1\}$ of independent $N(0,1)$ random variables and a sequence $\{f_n;n\ge 1\}$ of functions defined on $T\times B'$ so that the process $\overline{X}=\{\overline{X}(t,y');(t,y')\in T\times B'\}$ defined by:

$$\overline{X}(t,y') = \limsup_{N\to\infty} \sum_{n=1}^{N} \xi_n f_n(t,y') \tag{3.3}$$

is Gaussian, $\mathcal{T}\times\mathcal{B}_{\sigma(B')}$ measurable, $d_{\overline{X}}$-separable and its covariance is given by (3.1) (see for example [5.Théorème 3.1.4]). In fact for each (t,y') in $T\times B'$ the series in (3.3) converges almost surely. In the present situation, for each fixed $t\in T$, assumption (3.2) forces uniform convergence in $y'\in U$. Moreover, the functions f_n can be chosen in the reproducing kernel Hilbert space in order to be $\mathcal{T}\times\mathcal{B}_{\sigma(B')}$ measurable and linear and weak*continuous in y' for each fixed $t\in T$. Now for each $t\in T$ we define $X(t,\omega)$ as the unique element of B which, as a linear form on B', coincide with the right hand side of (3.3) if ω is such that the series converges uniformly in $y'\in U$, and 0 if not. $X=\{X(t);t\in T\}$ is a measurable B-valued Gaussian process which satisfies (1.1). Finally, for each fixed $t\in T$, we can use [5.Théorème2.1.2] and assumption (3.2) to show:

$$cst(U)E\{\|X(t)\|^2\} \le \varphi(t)^2\ E\{(\sup_{(u,u')\in U\times U} Z(u)-Z(u'))^2\}$$

and we conclude using the above Proposition ■

(5) $\mathcal{B}_{\sigma(B')}$ denotes the σ-field of subsets of B' generated by the elements of B viewed as linear forms on B'.

Let us give a new proof of claim (iv) of Section II.

Let (i,H,B) and $(j,K,L^p(\mathbf{T},\mathcal{T},\lambda))$ be two abstract Wiener spaces and let us denote by ν and μ the σ-additive extensions of the cylindrical measures $i(\gamma_H)$ and $j(\gamma_K)$. μ is the distribution of a measurable Gaussian process $\xi=\{\xi(t);t\in T\}$ the paths of which almost surely belong to $L^p(\mathbf{T},\mathcal{T},\lambda)$ and the covariance of which, say $K=\{K(s,t);(s,t)\in T\times T\}$ satisfies [7]:

$$\int_T K(t,t)^{p/2}\, d\lambda(t) < \infty.$$

Let us set:

$$\Gamma_{(s,t)}(x',y') = K(s,t)\int_B \langle x',x\rangle\langle y',x\rangle\, d\nu(x),$$

for s and t in T and x' and y' in B'. All the assumptions of our theorem are clearly satisfied with $\varphi(t) = K(t,t)^{1/2}$ and Z equals the restriction to any closed ball U of the Gaussian linear process canonically associated to ν (see for example [2. p.107]). The distribution of the process we obtain from the above theorem is then a Gaussian measure on $L^p(\mathbf{T},\mathcal{T},\lambda;\mathbf{B})$ which is easily seen to be the desired σ-additive extension of the cylindrical measure $i\otimes j(\gamma_{H\hat{\otimes}_2 K})$.

Possible extension

Let us first look at an example.
Let $X=\{X_p;p\geq 2\}$ be the $L^2([0,1])$-valued Gaussian process defined by:

$$X_p(t) = \sum_{n=1}^{\infty} a_n^{(p)} e_n(t)\, \xi_n$$

where $\{e_n;n\geq 1\}$ is the Haar system defined by:

$$e_{2^k+1}(t) = \begin{cases} 2^k & \text{if } t\in[(l-1)2^{-k},(2l-1)2^{-(k+1)}[\\ -2^k & \text{if } t\in[(2l-1)2^{-(k+1)},l2^{-k}[\\ 0 & \text{otherwise} \end{cases}, \quad k=1,2,\ldots \quad l=1,\ldots,2^k,$$

where $\{\xi_n;\ n\geq 1\}$ is a sequence of independent $N(0,1)$ random variables and where the coefficients $a_{2^k+1}^{(p)}$ are, for each fixed k, all zero except for exactly one $l=l(k)$ which is chosen so that the intervals $[(l(k)-1)2^{-k},l(k)2^{-k}[$ do not overlap, and

$$a_{2^k+1(k)}^{(p)} = k^{-a/p}\, 2^{-k(p-2)/2p}$$

for some real number $a>1$ independent of k. It is easy to check that for each $p\geq 2$ we have:

$$P\{X_p\in L^p([0,1])\} = 1,$$

and for each $\varepsilon>0$:

$$P\{X_p\in L^{p+\varepsilon}([0,1])\} = 0.$$

Consequently, interesting informations on the support of the distribution of the process are lost when looking at the process as at an $L^2([0,1])$-valued one. We should better look at the sample paths as elements of $\prod_{p\geq 2} L^p([0,1])$ rather than elements of $\prod_{p\geq 2} L^1([0,1])$.Viewing the paths as vector fields rather than functions should serve as a motivation for the study of Gaussian measures on "continuous sums of Banach spaces".

REFERENCES

[1]- A.BADRIKIAN: Prolégomènes au calcul des probabilités dans les Banach Lecture Notes in Math. # 539 (1976) 1-166

[2]- R.CARMONA: Tensor Product of Gaussian Measures Lecture Notes in Math. # 644 (1978) 96-124

[3]- R.CARMONA and S.CHEVET: Tensor Gaussian Measures on $L^p(E)$ J. Functional Anal. (to appear)

[4]- S.CHEVET: Un résultat sur les mesures gaussiennes C.R.Acad.Sc.Paris ser.A 284 (1977) 441-444

[5]- X.FERNIQUE: Régularité des trajectoires des fonctions aléatoires gaussiennes Lecture Notes in Math. # 480 (1975) 1-96

[6]- L.GROSS: Abstract Wiener Spaces. Proc. Fifth Berk. Symp. Math. Stat. and Proba. 2 (1965) 31-42

[7]- B.S.RAJPUT: Gaussian Measures on L_p spaces ($1\leq p<\infty$) J. Multivariate Anal. 2 (1972) 382-403

THE RADON-NIKODYM PROPERTY

S.D. Chatterji

§0. Introduction.

If E is a topological vector space then corresponding to each type (say (β)) of integration theory for functions taking values in E and each class of E-valued set-functions satisfying a certain set of conditions (say (α)), one can define a corresponding Radon-Nikodym property (RNP). For example, one could say that E has the α-β-RNP if for any probability space (Ω,Σ,P) and any E-valued set-function $\mu:\Sigma\to E$ satisfying (α) such that $\mu \ll P$ (i.e. $P(A) = 0$ implies $\mu(A) = 0$), one has $\mu(A) = \int_A f\, dP$, $\forall A \in \Sigma$, where f is P-integrable in the sense of (β). The most studied of these RNP's is the one where E is a Banach space, (β) is the strong (or Bochner) integral and (α) is the property that μ be σ-additive and of bounded variation (b.v.). For a Banach space E, it is this property that is called RNP. The factors contributing to its great success are two-fold; firstly, the general fact that strong integrals are more amenable and secondly, the very striking geometric characterisations and properties of this RNP discovered in the last few years (see below). However, the strong integral is a rarity in applications to operator theory and other areas and hence the other types of RNP need to be studied (and have been studied; see for references [2]).

RNP for a space E can be studied by various means. One of these is to use the scalar Radon-Nikodym theorem and combine it with lifting and some functional analytic juggling. What can be done easily by such methods is indicated in §2. Another method is to consider the ratios $\mu(A)/P(A)$ and pass to the limit as A becomes "small". If the space Ω is a subset of $\mathbb{R}^n$, it is clear what "small" A means; actually, it is sufficient for RNP to consider the case $\Omega = [0, 1]$ with P = Lebesgue measure (cf. Chatterji [3(i)]). Another way to take limits of $\mu(A)/P(A)$ is to use martingales and we make some comments on this in §3.

The present paper is entirely expository. Perhaps theorem 1 of §2, although very simple, is new and may be useful. In §3 we have tried to show that a combination of martingale theory and functional analysis gives an easy access to several theorems on RNP.

Our list of references is very short. A considerable bibliography is given in [1,2,4] . Results are cited by names of authors whose work can then be looked up in these references. Huff's articles [5] and in [1] give a clear exposition of dentability in relation to RNP. For vector-valued martingale theory we refer to [3] and on general vector-valued integration theory to [4].

§1. Notation.

In the sequel, we shall only consider the case of a Banach space E. If E' is its topological dual, the pairing will be denoted by $\langle x, x' \rangle = x'(x)$. (Ω,Σ,P) will always be the underlying probability space, $\Sigma^+ = \{A \mid P(A) > 0\}$, L^p, L^p_E will be the usual spaces of scalar or E-valued strongly measurable functions on (Ω,Σ,P). A simple function is one whose range is a finite set. By $f \cdot P$ we shall denote a set-function μ (scalar or E-valued) such that $\mu(A) = \int_A f\, dP$ for some suitable notion of integrability.

§2. RNP via scalar theory.

Let E be a Banach space and μ an E-valued σ-additive set-function of bounded variation defined on the σ-algebra Σ of a probability space (Ω,Σ,P). We may, without loss of generality, suppose that

$$||\mu(A)|| \leq M \cdot P(A) \ , \ \forall A \in \Sigma \qquad \ldots\ldots (1)$$

This is so since $||\mu(A)|| \leq \nu(A)$ where ν is the total variation of μ and $\nu << P$; thus (1) would be satisfied for a suitable probability measure Q such that $Q << P$ whence if $\mu = f \cdot Q$ then

$\mu = (f \cdot \frac{dQ}{dP}) \cdot P$. Another way to look at this is that if $A_n = \{\frac{d\nu}{dP} \leq n\}$

then $\mu_n(A) = \mu(A \cap A_n)$ is a vector measure satisfying (1) with $M = n$. Hence if $\mu_n = f_n \cdot P$ for all n then $f_n \to f$ a.e.(P) and $\mu = f \cdot P$.

Now for every $x' \in E'$ (the topological dual Banach space of E), $(x'\mu)(A) = \langle \mu(A), x' \rangle$ defines a scalar-valued measure $x'\mu << P$. Hence

$x'\mu = f_{x'} \cdot P$ where $f_{x'} \in L^{\infty}(\Omega,\Sigma,P)$. Clearly $x' \to f_{x'}$ is a linear map and

$$||f_{x'}||_{\infty} \leq M \cdot ||x'|| \ .$$

Hence we can choose a version of $f_{x'}$ by lifting theory (cf [8]) such that for all $\omega \in \Omega$ $x' \to f_{x'}(\omega)$ is linear and $|f_{x'}(\omega)| \leq M \cdot ||x'||$. This implies that $f_{x'}(\omega) = \langle x', f(\omega)\rangle$ where $f(\omega) \in E''$ i.e. $f : \Omega \to E''$ is E'-scalarly measurable, bounded (indeed $||f(\omega)|| \leq M$) and

$$\langle\mu(A), x'\rangle = \int_A \langle x', f\rangle \, dP$$

i.e.

$$\mu(A) = \int_A f \, dP$$

where the integral on the right hand side is taken in the E'-scalar sense. What we have thus shown is that any E-valued σ-additive measure μ of bounded variation can be written as $\mu = f \cdot P$ where $f : \Omega \to E''$ and integration is taken to be E'-scalar integration.

The whole problem of RNP is then to see under what conditions on E or μ, the map $f : \Omega \to E''$ has its values in E a.s. and that its range is a.s. separable. If E is separable and reflexive we see therefore that E will have RNP. However, since for RNP it suffices to consider $\Omega = [0, 1]$, in which case the range of μ is automatically separable, it is clear that a space E has RNP if and only if all its (closed) separable subspaces have RNP. Since every subspace of a reflexive space is reflexive we see that any reflexive space has RNP.

If we repeat the above argument with μ taking values in a space E' but now consider only the scalar measures $x\mu$, $x \in E$ (where $(x\mu)(A) = \langle x, \mu(A)\rangle$) we obtain that $x\mu = f_x \cdot P$ where $x \to f_x(\omega)$ is linear and $|f_x(\omega)| \leq M \cdot ||x||$ for all $\omega \in \Omega$. Thus there exists $f : \Omega \to E'$, E-scalarly measurable and such that $f_x(\omega) = \langle x, f(\omega)\rangle$ whence

$$\mu(A) = \int_A f \, dP \ , \quad A \in \Sigma$$

where the integral on the right is taken in the E-scalar sense. We see therefore that any E'-valued σ-additive set-function of bounded variation $\mu \ll P$ has an integral representation. Thus E' will have RNP if and only if every $f : \Omega \to E'$, $||f(\omega)|| \leq M$, and E-scalarly integrable is actually strongly measurable. If E' is separable (so that E is also separable), E-scalar measurability implies strong measurability; this is easy and follows from a lemma below in §3. Thus if E' is separable, E' has RNP.

For a general E', it follows from our discussion concerning E that E' will also have RNP (even if it is non-separable) provided that every separable subspace of E has a separable dual. This observation (due to Uhl) follows from noting that if S is a separable subspace of E' then there exists a separable subspace F of E such that S is isomorphic to a subspace of F' and hence S will have RNP; this being true for any separable subspace S of E', E' has RNP. A striking theorem due to Stegall completes this line of investigation; it says that if E' has RNP then the dual of any separable subspace of E is also separable. One proof would consist in showing that if E is separable but E' is not then there exists $f : \Omega \to E'$, $||f(\omega)|| \leq M$. f E-scalarly measurable but not having separable range a.s. Then the set-function $\mu(A) = \int_A f \, dP \in E'$ has all the prescribed properties but is not representable by a strongly integrable integrand so that E' fails to have RNP. Such a proof can be given following Stegall's construction in [7].

Thus RNP for E' is completely characterised. As for RNP for a Banach space E, the key geometric condition is that of dentability. A set A in E is called dentable if for any $\varepsilon > 0$ there is $x \in A$ (where x may depend on ε) such that x is not in $\overline{co}\{A \setminus B(x,\varepsilon)\}$ where $B(x,\varepsilon) = \{y : ||y-x|| < \varepsilon\}$ and $\overline{co}$ stands for the norm closure of the convex hull. A Banach space E has RNP if and only if every closed bounded convey subset of E is dentable. The "if" part of the theorem (as well as the notion of dentability) is due to Rieffel [6] and the converse is due to Davis and Phelps and independently to Huff. We shall comment on the proof of this theorem via martingale theory in §3.

A number of geometric properties of dentable sets have been obtained by Davis, Phelps, Huff, Morris, Bourgain, Lindenstrauss,

Namioka, Edgar, Weizsäcker, Stegall and others. Let us simply mention some very easy facts concerning dentability. If A is relatively norm compact then A is dentable ([6]); a more difficult result which follows indirectly through the considerations of §3 is that any relatively weakly compact set A is dentable. Also if $\overline{co}(A)$ is dentable so is A ([6]). In $\mathbb{R}$, a subset A is dentable if and only if either inf A > $-\infty$ or sup A $<\infty$. In $\mathbb{R}^n$, $n \geq 2$, any half-space is non-dentable. It would appear that dentability is most interesting in the case of closed, bounded, convex sets.

Another class of spaces E for which RNP holds is that of spaces having a boundedly complete Schauder basis i.e. E has a Schauder basis $\{x_n\}$ (i.e. any $x \in E$ can be written uniquely as $\sum_n a_n x_n$) such that if $\sup_N \|\sum_{n=1}^N a_n x_n\| < \infty$ then $\sum_{n=1}^\infty a_n x_n$ exists. This was proved very early by Dunford and Morse. The following theorem generalises this result in a very simple way. Whether any really new and interesting class of spaces fall within the ambit of the theorem is unknown to me.

<u>Theorem 1</u>.

Let E be a Banach space such that there exists a sequence of bounded linear operators $T_n : E \to E_n$ with the following properties:
(i) each E_n is a closed subspace of E and has RNP ;
(ii) $T_n T_{n+1} = T_{n+1} T_n = T_n$; (iii) $\lim_{n\to\infty} T_n x = x$ for all $x \in E$;
(iv) if $\{y_n\}$ is a bounded sequence in E such that $T_n y_{n+1} = y_n$ then $\lim_n y_n$ exists. Then E has RNP.

<u>Proof</u>:

Let (Ω,Σ,P) be a probability space and $\mu : \Sigma \to E$ be an additive set-function such that $\|\mu(A)\| \leq M \cdot P(A)$. Then (from (i) and (iii))

$$\mu(A) = \lim_{n\to\infty} T_n(\mu(A)) = \lim_n \int_A f_n \, dP$$

where $f_n(\omega) \in E_n$ and $\|f_n(\omega)\| \leq M$. Now (from (ii))

$$T_n(\mu(A)) = T_n(T_{n+1}\,\mu(A)) = T_n\left(\int_A f_{n+1}\,dP\right) = \int_A T_n\, f_{n+1}\,dP$$

which gives $f_n(\omega) = T_n(f_{n+1}(\omega))$ a.e. Because of (iv) $\lim_n f_n(\omega) = f(\omega)$ exists a.e. and we deduce that $\mu(A) = \int_A f\,dP$; since each f_n is strongly measurable so is f and the theorem is proved.

Remark: In case E has a boundedly complete Schauder basis $\{x_n\}$, we take E_n to be the subspace spanned by $\{x_1, \ldots, x_n\}$ and $T_n\, x = \sum_{k=1}^{n} a_k\, x_k$ if $x = \sum_1^\infty a_n\, x_n$. Obviously (i), (ii) and (iii) are satisfied. The bounded completeness of the basis $\{x_n\}$ implies (iv) since any sequence $\{y_n\}$ such that $T_n\, y_{n+1} = y_n$ must be of the type $y_n = \sum_{j=1}^{n} a_j\, x_j$.

We close this section by remarking that if E is separable the lifting argument used in the beginning can be avoided by a suitable choice of the densities f_x. This elementary reasoning was first used by Dieudonné; the explosion of lifting theory in the 60's made this type of proof via lifting very popular.

Spaces which do not have RNP are typically as follows: c_0 , ℓ^∞, $L^\infty[0,1]$, $C[0,1]$, $L^1[0,1]$; this can be verfied easily either by methods of this section or those of §3. However, much more subtle examples are known too (cf. [4(ii)] for references).

§3. RNP via martingales.

The basic idea behind the martingale approach to RNP is quite old and was used right from the very beginning of differentiation theory. It was fully crystallised in our article [3(i)] and it is our fond hope that the methods used there will continue to be of some use. In modern notation, the method is as follows. If $\mu : \Sigma \to E$ is an additive set-function on some probability space (Ω,Σ,P) then to any denumerable partition $\pi = \{A_i\}$ of Ω with $P(A_i) > 0$ (we write: $A_i \in \Sigma^+$) we can associate the function $f_\pi = \sum_i x_i \phi_{A_i}$ where $x_i = \mu(A_i) / P(A_i)$. If π's are ordered in the obvious way (i.e. $\pi_1 \leqslant \pi_2$ iff each set of the partition π_1 is a.s. the union of sets of the partition π_2) then $\{f_\pi\}$ forms an E-valued martingale. If $||\mu(A)|| \leqslant M \cdot P(A)$ then $||f_\pi(\omega)|| \leqslant M$. Then it is easily seen that μ will be of the form $f \cdot P$ with strongly measurable f and only if $\int ||f_\pi - f|| \, dP \to 0$ as $\pi \to \infty$. Since this is metric convergence, it suffices to verify the L_E^1-convergence of every sequence $g_n = f_{\pi_n}$ with $\pi_1 \leqslant \pi_2 \leqslant \ldots$; since $||g_n(\omega)|| \leqslant M$, it is enough to have convergence a.s. or in probability. It is shown in [3(i)] that any general martingale convergence theorem in E is equivalent to RNP for E. In particular E will have RNP if and only if every uniformly bounded simple (even two-valued) martingale $\{g_n\}_{n \geqslant 1}$ converges a.s. From this we can now see how a proof of the dentability characterisation of RNP (see §2) will go. If E is dentable i.e. every bounded closed convex subset of E is dentable, then so is any bounded subset of E. Now the set $S = \{\mu(B) / P(B) \mid B \in \Sigma^+\}$ is a bounded set if $||\mu(B)|| \leqslant M \cdot P(B)$ and hence dentable. Using dentability one shows that if $A \in \Sigma^+$, $\varepsilon > 0$, there is a subset $A_0 \in \Sigma^+$ of A such that the diameter of the set $\{\mu(B) / P(B) \mid B \subset A_0, B \in \Sigma^+\}$ is smaller than ε. This is the crux of Rieffel's argument ([6]); I shall not repeat it here although it is quite short (cf. e.g. Huff's article in [1]). Once this is done, it is clear via a standard exhaustion argument that given any $\varepsilon > 0$ there is a denumerable partition π_ε such that $||f_\pi(\omega) - f_{\pi'}(\omega)|| < \varepsilon$ for $\pi \geqslant \pi_\varepsilon$, $\pi' \geqslant \pi_\varepsilon$. Thus $f_\pi \to f$ a.s. and in L_E^1 whence $\mu = f \cdot P$. Conversely to prove that RNP implies dentability, we proceed by contradiction. If S is a bounded closed convex <u>non-dentable</u> set, we shall construct a simple martingale $\{\xi_n\}$ with values in S such that $\lim_n \xi_n$ does not exist and hence E will not have RNP. The argument is essentially due to Huff (following perhaps a construction due to Maynard). We choose $\Omega = [0, 1]$, P = Lebesgue measure and define $f_0(\omega) \equiv x_{00} \in S$ where $x_{00} = x$ is an arbitrary point in S. Because of the non-dentability of S, there exists an $\varepsilon > 0$ such that for any $x \in S$, $x \in \overline{co}\,\{S \setminus B(x,\varepsilon)\}$; in particular, we can find $x_1, \ldots, x_n$ in S such that $||x_j - x|| \geq \varepsilon$, $j \geq 1$, but $||x - \sum_{j=1}^{n} \alpha_j x_j|| < \frac{1}{2}$ where $\alpha_j > 0$, $\sum_j \alpha_j = 1$. By repeating some of the x_j's we may suppose that $0 < \alpha_j \leq \frac{1}{2}$. If we now define $f_1 = \sum_j x_j \phi_{I_i}$ where I_j are intervals of length α_j we see that $||\mathbb{E}_0 f_1(\omega) - f_0(\omega)|| < \frac{1}{2}$ where E_0 is the conditional expectation operator given the sub-field generated by f_0 (in this case trivial). Continuing this process, we define a sequence $\{f_n\}$ of E-valued functions such that $f_n(\omega) \in S$, $||f_{n+1}(\omega) - f_n(\omega)|| \geq \varepsilon$ but

$$||(E_n f_{n+1})(\omega) - f_n(\omega)|| \leq \tfrac{1}{2}^{n+1} \qquad \ldots\ldots(*)$$

where E_n is the conditional expectation operator given $f_0, f_1, \ldots, f_n$. If we write

$$g_n = f_0 + \sum_{j=1}^{n} (f_j - E_{j-1} f_j)$$

then $\{g_n\}$ is a martingale. Since

$$f_n = g_n + \sum_{j=1}^{n} (E_{j-1} f_j - f_{j-1})$$

and $\sum_{j=1}^{\infty} ||E_{j-1} f_j - f_{j-1}|| \leq \sum_j 2^{-j} < \infty$, $\lim_n g_n(\omega)$ exists if and only if $\lim_n f_n(\omega)$ exists; but the latter exists nowhere so the martingale $\{g_n\}$, although uniformly bounded, does not converge anywhere. This is enough to show that E does not have RNP but the values of g_n are not quite in S; by starting the martingale at a large n_0 we shall have the values in some δ-neighbourhood of S. To define a non-convergent martingale

$\{\xi_n\}$ with values actually in S, we may put

$$\xi_n(\omega) = \lim_{N\to\infty} E_n f_N(\omega).$$

Then $\xi_n(\omega) \in S$ is well-defined and is a martingale; indeed

$$||E_n f_{N'}(\omega) - E_n f_N(\omega)|| \leqslant 2^{-N}$$

for $N' > N \geqslant n$ as a standard calculation using (*) will show. Hence $||\xi_n(\omega) - f_n(\omega)|| \leqslant 2^{-n}$ and the impossibility of the existence of $\lim_n \xi_n(\omega)$ anywhere. (I learnt the last part of the argument from some notes of Garling.)

We now illustrate another martingale argument to prove a RNP proposition for E' due to Stegall ([7] Corollary 5). This type of argument was used in [3(i)] to prove Phillip's Radon-Nikodym theorem for weakly compact subsets. We prove that if E is a Banach space such that any bounded sequence in E' has a $\sigma(E', E)$-closure which is norm separable in E' then E' has RNP.

For the proof of this, we need only show that any E'-valued, simple, uniformly bounded martingale $\{f_n\}$ converges a.s. in E'. Since $\{f_n(\omega) \mid n \geqslant 1, \omega \in \Omega\}$ is denumerable, by hypothesis, its $\sigma(E', E)$-closure T is norm separable in E'. But T, being bounded, is $\sigma(E', E)$-compact; so we can define $\bar{f}(\omega) = \lim f_n(\omega) \in T$ by taking the limit following an ultrafilter finer than the Cauchy filter on the integers. On the other hand, $\omega \mapsto \langle x, f_n(\omega)\rangle$ is a scalar valued uniformly bounded martingale and hence convergent a.s. This tells us that $\lim_{n\to\infty} \langle x, f_n(\omega)\rangle = \langle x, \bar{f}(\omega)\rangle$ a.s. whence $\bar{f}$ is seen to be E-scalarly measurable with respect to the completed σ-algebra $\bar{\Sigma}$ of the basic probability space (Ω,Σ,P). By the lemma which follows, there exists a Σ-strongly measurable f such that $\lim_{n\to\infty} \langle x, f_n(\omega)\rangle = \langle x, f(\omega)\rangle$ a.s. Because of the uniform boundedness of the f_n's we conclude that for any $x \in E$,

$$\int_A \langle x, f \rangle \, dP = \lim_{n\to\infty} \int_A \langle x, f_n \rangle \, dP$$

$$= \int_A \langle x, f_N \rangle \, dP$$

if $A \in \Sigma_N$ - σ-algebra generated by the f_j, $j \leq N$. But then this implies that $f_N = E(f|\Sigma_N)$ whence we conclude the existence a.s. of $\lim_{N\to\infty} f_N(\omega)$. We now give the needed

Lemma:

Let $\overline{f} : \Omega \to E'$ be E-scarlarly measurable with respect to $\overline{\Sigma}$, the P-completion of Σ. If E is a separable Banach space and $\overline{f}$ is almost separably valued then $\overline{f}$ is strongly $\overline{\Sigma}$-measurable. Hence there is a strongly Σ-measurable E'-valued function f such that $f = \overline{f}$ a.s.

Proof:

It is enough to prove the statement concerning $\overline{\Sigma}$-measurability of $\overline{f}$; the existence of f follows easily. Since $\overline{f}$ is almost separably valued we need only consider the case where the values of $\overline{f}$ lie in a separable supspace S of E'. Now $||\overline{f}(\omega) - a'|| = \sup_i |\langle x_i, \overline{f}(\omega) - a'\rangle|$ for a suitable choice of the denumerable set $x_i \in E$ and any $a' \in E'$. This means that the inverse image of any ball in S is $\overline{\Sigma}$-measurable. S being separable, its Borel sets form exactly the smallest σ-algebra generated by its balls. This implies the strong $\overline{\Sigma}$-measurability of $\overline{f}$.

§4. Conclusion.

The notion of RNP can be localised to a subset A of E. We say that A has RNP if any E-valued measure μ such that $\mu(B) / P(B)$ is in A for any set B has representation $f \cdot P$. If we say that A has the martingale convergence property (MCP) if any A-valued martingale converges a.s. then the discussion of §3 shows that for bounded, closed, convex sets A, dentability, RNP and MCP are equivalent conditions.

Many other probabilistic aspects of RNP have been ignored in this article. We mention here at least the work on vector-valued amarts due to Bellow, Brunel, Chacon, Edgar, Sucheston and others and the work of Assouad, Enflo, James, Pisier and others on super-reflexivity, super-RNP and uniform convexity.

References.

[1] Measure theory, Oberwolfach 1975. Lecture Notes in Mathematics, No. 541; Springer-Verlag, Berlin (1976). Ed. A. Bellow and D. Kölzow.

[2] Vector space measures and applications (I,II). Proceedings, Dublin 1977. Lecture Notes in Mathematics, Nos. 644, 645; Springer-Verlag, Berlin (1978). Ed. R.M. Aron and S. Dineen.

[3] Chatterji, S.D.
(i) Martingale convergence and the Radon-Nikodym theorem. Math. Scand. 22, 21-41 (1968).
(ii) Vector-valued martingales and their applications, in Probability in Banach spaces, Oberwolfach 1975. Lecture Notes in Mathematics, No. 526; Springer-Verlag, Berlin (1976). Ed. A. Beck.

[4] Diestel, J., Uhl, J.J. Jr.
(i) The Radon-Nikodym theorem for Banach space valued measures. Rocky Mountain Math. J. 6, 1-46 (1976).
(ii) Vector measures. AMS-Mathematical surveys 15 (1977).

[5] Huff, R.E.
The Radon-Nikodym property for Banach spaces - a survey of geometric aspects, in Functional analysis: Surveys and recent results, proceedings of the Paderborn conference on functional analysis (Ed. K.-D. Bierstedt and B. Fuchssteiner). Notas de Matemática (63), Mathematics Studies, N.-Holland Amsterdam (1977).

[6] Rieffel, M.A.
Dentable subsets of Banach spaces, with applications to a Radon-Nikodym theorem, Proc. Conf. Functional Analysis, Thompson Book Co., Washington D.C. 71-77 (1967).

[7] Stegall, C.
The Radon-Nikodym property in conjugate Banach spaces. Trans. Amer. Math. Soc. 206, 213-223 (1975).

[8] Tulcea, A. Ionescu and C. Ionescu
Topics in the theory of lifting. Springer-Verlag, Berlin (1969).

S.D. Chatterji
Dépt. de Mathématiques
Ecole Polytechnique Fédérale
de Lausanne
61, av. de Cour
1007 Lausanne / Switzerland

On B-convex Orlicz Spaces

Manfred Denker and Rainer Kombrink
(Institut für Mathematische Statistik, University of Göttingen)

It is well known that L_p- and l_p-spaces are B-convex if and only if they are reflexive. Though one would suspect that B-convex Orlicz spaces are precisely the reflexive ones, for various reasons there has been made no attempt so far in proving a statement like that. We study the question which Orlicz spaces are B-convex and list some equivalent conditions due to various authors.

It seems that there has not been paid much attention to the theory of Orlicz spaces (cf. Lindenstrauss-Tzafriri [8]). In 1977, however, Lindenstrauss and Tzafriri ([9]) gave a first detailed description of Orlicz sequence spaces taking into account their work ([10]) about the theory of subspaces of such Orlicz spaces, from which it is almost obvious to derive the classification of B-convex Orlicz sequence spaces. Therefore, we first describe the facts and theorems from which our theorem for sequence spaces follows filling in some facts which are not difficult to prove. Because Akimovič's paper [1] seems to be almost unknown we shall also sketch the proof of his result. For general Orlicz spaces some more efforts have to be made in order to derive the same equivalences. For this reason we start with sequence spaces and then consider arbitrary Orlicz spaces.

An Orlicz function M is a nonnegative, non-decreasing, convex and continuous function defined on $\mathbb{R}_+$, such that $M(0) = 0$ and $\lim_{x \to \infty} M(x) = +\infty$. (Later we have to use a slightly more general definition.) We shall say that an Orlicz function M satisfies the δ_2-condition if there exist an $x_0 > 0$ and a $K > 0$ such that $M(2x) \leq KM(x)$ for each $0 \leq x \leq x_0$. Every Orlicz function M has a derivative m (from the right), and by [6] a representation $M(x) = \int_0^x m(t)dt$. If m satisfies $m(0) = 0$ and $m(x) > 0$ for $x > 0$, then M is said to be a proper Orlicz function. Finally for such functions M we can define its complementary Orlicz function N by $N(y) = \int_0^y n(s)ds$, where $n(s) = \sup\{t \mid m(t) \leq s\}$ denotes the inverse function of the derivative m of M .

The Orlicz sequence space l_M with respect to the Orlicz function M is the set of all real valued sequences (x_n) for which there exists a $t > 0$ with $\sum M(t|x_n|) < \infty$. l_M is a Banach space with the norm $\|(x_n)\|_M = \inf\{t > 0 \mid \sum M(t^{-1}|x_n|) \leq 1\}$. An important role plays the closed, separable subspace

$h_M = \{(x_n) \in l_M \mid \sum M(t|x_n|) < \infty$ for all $t \geq 0\}$,

of which we shall make use in the following.

For our first theorem we need a few facts, which can be found in Gribanov [4], Lindberg [7], Lindenstrauss-Tzafriri [9] and Luxemburg [11]:

(1) The following are equivalent:

a) M satisfies the δ_2-condition

b) $l_M = h_M$

c) l_M is separable

(2) If M and N are complementary Orlicz functions, then l_N is seperable iff h_M does not contain a complemented subspace isomorphic to l_1

(3) If M and N are complementary Orlicz functions, then l_N is isomorphic to the dual space h_M^* of h_M.

(4) If $M(x) = 0$ for some $x > 0$, then l_M is isomorphic to l_∞, and if $m(0) > 0$, then l_M is isomorphic to l_1.

<u>Theorem 1</u>: For Orlicz sequence spaces the following are equivalent:

(1) M is a proper Orlicz function and M and its complementary function N satisfy the δ_2-condition

(2) l_M is B-convex

(3) l_M is uniformly convexifiable.

<u>Proof</u>: We show "(3) ⇒ (2) ⇒ (1) ⇒ (3)".

"(3) ⇒ (2)" has been remarked by Beck [2].

For proving "(2) ⇒ (1)" let l_M be B-convex. Then l_M is not isomorphic to l_1 or l_∞ and thus by fact (4) above, M is a proper Orlicz function and N is well defined. Moreover, since h_M is B-convex, h_M does not contain a (complemented) subspace isomorphic to l_1 and therefore l_N is separable (by (2)) and N satisfies the δ_2-condition (by (1)). Also by (1), $l_N = h_N$ and, by (3), h_M^* is isomorphic to l_N. A theorem of Gisey ([5]) says that a Banach space is B-convex iff its dual is, so that l_N is B-convex. Repeating the same arguments, we obtain the δ_2-condition for M.

The implication "(1) ⇒ (3)" follows from the work of Akimovič [1], Lindberg [7] and Luxemburg [11], which we shall repeat here briefly as mentioned above. It is well known that two equivalent Orlicz functions M_1

and M_2 (i.e.: there exist positive constants K_1, k_1, K_2, k_2 and $x_o > 0$ with $K_1 M_1(k_1 x) \leq M_2(x) \leq K_2 M_1(k_2 x)$ for $x \leq x_o$) generate equivalent norms. Thus, it has to be shown that there exists an equivalent $\bar{M}$ generating a uniformly convex norm.

Let M and N be complementary Orlicz functions, both satisfying the δ_2-condition. We may assume that M has a continuous, strictly increasing derivative m, linear for x larger than some x_o, and that there exists a constant $K(> 2)$ satisfying $M(2x) \leq K\,M(x)$ and $N(2x) \leq K\,N(x)$ for every $x \geq 0$ (see [7]). For otherwise take a function M_1, equivalent to M, defined by $M_1(x) = \int_o^x \frac{M(t)}{t}\,dt$ $(x \leq x_o)$ and by $M_1(x) = M_1(x_o) + \int_{x_o}^x x_o^{-1} M(x_o) t\,dt$ $(x \geq x_o)$, where x_o is small enough depending on the δ_2-condition of M and N. Then to prove that N_1 satisfies the δ_2-condition for every $x \geq 0$, use the fact from [6] that for equivalent pairs of complementary Orlicz functions (M_o, N_o) and (M_o', N_o') with $M_o(x) \leq M_o'(x)$ $(x \leq x_o$, resp. $x \in \mathbb{R}_+)$ one has $N_o'(y) \leq N_o(y)$ $(y \leq y_o$ for some $y_o > 0$, resp. $y \in \mathbb{R}_+)$ and use $M(2x) \geq K'M(x)$ for some $K' > 1$, because N satisfies the δ_2-condition.

The essential part of the construction is due to Akimovič [1]: There exists an equivalent function $\bar{M}$ with derivative $\bar{m}$ satisfying the δ_2-condition for $x \in \mathbb{R}_+$, such that for each $e > 0$ there exists a $k = k(e) > 1$ with $\bar{m}((1+e)x) \geq k\,\bar{m}(x)$ for every $x \in \mathbb{R}_+$. This is shown as follows:

Let $e_o := (2^{-1}K)^{p+2} - 1$ where $p < {}_2\log K \leq p+1$, $c = 1 + e_o$ and $x_n = x_o c^{-n}$ $(n=0,1,2,\ldots)$, where x_o is as above. Define $\bar{m}(0) = 0$, $\bar{m}(x_n) = m(x_n)$, $\bar{m}(t) = m(t)$ $(t \geq x_o)$ and $\bar{m}$ on (x_{n+1}, x_n) by linear interpolation $(n=0,1,2,\ldots)$. Note that $\bar{M}(x) = \int_o^x \bar{m}(t)dt$ is equivalent to M, since for $x_{n+1} \leq t \leq x_n$ one has

$$m(c^{-1}t) = m\left(\frac{t}{c\,x_{n+1}}\,x_{n+1}\right) \leq m(x_{n+1}) \leq \bar{m}(t) \leq m(x_n) \leq m(ct),$$

and integration gives the equivalence of M and $\bar{M}$. Also $\bar{M}$ satisfies the δ_2-condition for every $x \geq 0$.

For completing the proof of the statement above, we only have to consider $0 < e \leq e_o$, since for $e \geq e_o$ $\bar{m}((1+e)x) \geq \bar{m}(cx)$. So, let $0 < e \leq e_o$ and set $d = 1+e$. If x and dx belong to the same interval $[x_{n+1}, x_n[$ $(n=-1,0,1,\ldots)$ where $x_{-1} = +\infty$, then it follows from the linearity of $\bar{m}$ on this interval by an easy calculation that (with $\bar{m}(t) = u + vt$ on $[x_{n+1}, x_n[$ and $u \geq 0$ for example)

$$\frac{\bar m(dx)}{\bar m(x)} \ge \frac{\bar m(d\,x_{n+1})}{\bar m(x_{n+1})} = \frac{v\,d\,x_{n+1}+u}{m(x_{n+1})} = \frac{m(x_{n+1}) + v(d-1)x_{n+1}}{m(x_{n+1})}$$

$$\begin{cases} = 1 + \dfrac{m(x_n)-m(x_{n+1})}{m(x_{n+1})}\;\dfrac{(d-1)x_{n+1}}{x_n-x_{n+1}} \ge 1 + (k_1-1)\,\dfrac{d-1}{e-1} & \text{for } n \ge o \\ = 1 + v(d-1)\,\dfrac{x_o}{m(x_o)} & \text{for } n=-1 \end{cases}$$

The other case, where x and dx belong to an interval of the form $[x_{n+1},x_{n-1}[$ $(n \ge o)$ is shown similarly. We should remark that k_1 is a constant independent of n and is used to estimate $m(x_n) = m(c\,x_{n+1}) \ge k_1\,m(x_{n+1})$. In order to show this, note that (M,N) and $(\tilde M(x) = K^{-1}M(2^{-1}Kx),\ \tilde N(y) = K^{-1}N(2y))$ are equivalent pairs of complementary Orlicz functions. Since $\tilde N(y) \le N(y)$ it follows from a remark above (cf. [6]) that $M(x) \le \tilde M(x) = K^{-1}\,M(2^{-1}Kx)$ and hence by the definition of c

$$m(cx) \ge \frac{M(cx)}{cx} \ge x^{-1}\,M(x)\,2^{p+2} \ge K^{-1}\,2^{p+2}m(x);$$

especially this holds for $x = x_{n+1}$ and $k_1 = K^{-1}2^{p+2} \ge 2$.

Next we show that for any $a \in (0,1)$ there exists a $d = d(a) \in (0,1)$ such that

$$(*)\qquad \bar M(\tfrac{x+ax}{2}) \le 2^{-1}(1-d)(\bar M(x) + \bar M(ax)) \qquad (x \in \mathbb{R}_+).$$

Indeed, with $b = \frac{x+ax}{2}$, we have the equivalent estimation

$$(2\,\bar M(b))^{-1}(\bar M(x)+\bar M(ax)) = 1 + (2\bar M(b))^{-1}\left(\int_b^x \bar m(t)dt - \int_{ax}^b \bar m(t)dt\right)$$

$$\ge 1 + (2\,\bar M(b))^{-1}\int_{ax}^b [\bar m(t + \tfrac{1-a}{1+a}\,t) - \bar m(t)]\,dt$$

$$\ge 1 + \frac{(k-1)\bar m(ax)(b-ax)}{2b\,\bar m(b)} \ge 1 + \frac{k-1}{2}\,\frac{1-a}{1+a}\,K^*(a).$$

The proof is now finished applying Theorem 1, p.64 of [11], which says that an Orlicz function satisfying (*) generates a uniformly convex norm.

q.e.d.

From now on we consider a general class of Orlicz spaces, as described below. First we extend the definition of an Orlicz function M, which, from now on, is understood to be a nonnegative, non-decreasing, convex and continuous function on some interval $[0,a[$ $(0 < a \le \infty)$ and $M(x) = \infty$ for $x \ge a$. (cf. [11]). (For sequence spaces this gives nothing new.) Let (Ω,F,μ) be a complete σ-finite measure space satisfying one of the following conditions:

(a) $0 < \mu(\Omega)$ and there exists a subset $\Omega_o \in F$ with $\mu(\Omega_o) = \mu(\Omega)$, such that μ has no atoms in Ω_o.

(b) $0 < \mu(\Omega) < \infty$ and μ is concentrated on countably many points $a_1, a_2, \dots$ such that $\mu(a_n) \geq \mu(a_{n+1})$ and

$$\liminf \frac{\mu(a_{n+1})}{\mu(a_n)} > 0.$$

(c) $\mu(\Omega) = \infty$ and μ is purely atomic with

$$0 < \liminf \mu(a_n) \leq \limsup \mu(a_n) < \infty.$$

In the following we always assume these restrictions.

We define the Orlicz function space $L_M(\mu) = L_M(\Omega, F, \mu)$, where M is an Orlicz function, by

$$L_M(\mu) = \{f \mid f : \Omega \to \mathbb{R} \text{ measurable and } \int M(t|f|)d\mu < \infty \text{ for some } t > 0\}$$

and the norm $\|\cdot\|_M$ by

$$\|f\|_M = \inf\{t > 0 \mid \int M(t^{-1}|f|)d\mu \leq 1\}.$$

An Orlicz function M is said to satisfy the Δ_2-condition if there exist positive constants K and $x_o < a$ such that $M(2x) \leq KM(x)$ for $x \geq x_o$. Considering the different underlying measure spaces, we say that an Orlicz function has the delta-property if

(α) M satisfies the Δ_2-condition, whenever $\mu(\Omega) < \infty$.

(β) M satisfies the δ_2- and the Δ_2-condition, whenever $\mu(\Omega) = \infty$ and μ has a non-atomic part of infinite measure (see (a) above).

(γ) M satisfies the δ_2-condition, whenever we are in case (c) above.

In the case of a finite measure μ, we can always assume that the Orlicz function is proper. For if M is an Orlicz function, for which $x_o = \sup\{x \geq 0 \mid M(x) = 0\} > 0$, define $M_1(x) = M(x_o+x)$ $(x \geq 0)$, and show that $L_M(\mu)$ and $L_{M_1}(\mu)$ are isomorphic. Indeed, M_1 is clearly an Orlicz function, for $0 \neq f \in L_{M_1}(\mu)$ one has

$$\int M\Big(\frac{|f|}{\|f\|_{M_1}}\Big)d\mu \leq \int M\Big(x_o + \frac{|f|}{\|f\|_{M_1}}\Big)d\mu \leq 1 ,$$

and for $0 \neq f \in L_M(\mu)$ it follows by convexity that

$$\int M_1\Big(\frac{|f|}{\|f\|_M}(1-a)\Big)d\mu = \int M\Big(a(a^{-1}x_o) + (1-a)\frac{|f|}{\|f\|_M}\Big)\,d\mu$$

$$\leq a \int M(a^{-1}x_o)d\mu + (1-a)\int M\Big(\frac{|f|}{\|f\|_M}\Big)d\mu \leq 1 ,$$

where $0 < a < 1$ is chosen so that $M(a^{-1}x_o)\mu(\Omega) \leq 1$.
Therefore, $L_M(\mu) = L_{M_1}(\mu)$ and $\|f\|_M \leq \|f\|_{M_1} \leq (1-a)^{-1}\|f\|_M$.

On the other hand, if M is an Orlicz function for which $m(0) > 0$, then it just has been shown that $M_1(x) = M(x-x_o)$ $(x \geq x_o)$, $M_1(x) = 0$ $(x \leq x_o)$ for some $x_o > 0$ generates an isomorphic space $L_{M_1}(\mu)$.

Choose $y > x_o$ with $M_1(y) = a$ $(a\,\mu(\Omega) \le 1/2)$ and define $M_2(x) = M_1(x)$ $(x \ge y)$ and by a continuous, decreasing, convex extension with $m_2(0) = 0$ on $[0,y[$, such that $M_2(x) \ge M_1(x)$. Then if $0 \neq f \in L_{M_2}(\mu)$

$$\int M_1\left(\frac{|f|}{\|f\|_{M_2}}\right)d\mu \le \int M_2\left(\frac{|f|}{\|f\|_{M_2}}\right)d\mu = 1$$

and if $0 \neq f \in L_{M_1}(\mu)$ then

$$\int M_2\left(\frac{|f|}{2\|f\|_{M_1}}\right)d\mu \le \int_{\{|f|<2y\|f\|_{M_1}\}} M_2\left(\frac{|f|}{2\|f\|_{M_1}}\right)d\mu +$$

$$+ \int_{\{|f|\ge 2y\|f\|_{M_1}\}} M_1\left(\frac{|f|}{2\|f\|_{M_1}}\right)d\mu$$

$$\le a\,\mu(\Omega) + 1/2 \int M_1\left(\frac{|f|}{\|f\|_{M_1}}\right)d\mu \le 1$$

Therefore, $L_{M_1}(\mu) = L_{M_2}(\mu)$ and $\|f\|_{M_1} \le \|f\|_{M_2} \le 2\|f\|_{M_1}$.

<u>Theorem 2</u>: For Orlicz function spaces $L_M(\mu)$, for which the measure satisfies one of the conditions (a) - (c), the following are equivalent:

(1) M is a proper Orlicz function and M and N have the delta-property

(2) $L_M(\mu)$ is B-convex

(3) $L_M(\mu)$ is uniformly convexifiable

<u>Proof</u>: "(3) $\Rightarrow$ (2)" follows from Beck [2] again, and "(1) $\Rightarrow$ (3)" is shown by a similar method as in the proof of theorem 1. This is also due to Akimovič [1] and Luxemburg [11]; a special case was done by Milnes [13].

We show now "(2) $\Rightarrow$ (1)".

Let $L_M(\mu)$ be a B-convex Orlicz space. By the remark above, if $\mu(\Omega) < \infty$ we may assume, that M is proper. If $\mu(\Omega) = \infty$, suppose that M is not a proper Orlicz function. This can fail for two reasons, either $M(x) = 0$ for some $x > 0$ or $m(0) > 0$.

Consider the first case. Let $x_o = \sup\{x \mid M(x) = 0\} > 0$, and let $\varepsilon > 0$ and $k \in \mathbb{N}$. Since $\mu(\Omega) = \infty$, there exist 2^k disjoint sets E_i satisfying $\inf M\left(\frac{x_o}{1-\varepsilon}\right)\mu(E_i) \ge 1$. Denote by $(\sigma_{ij})_{j=1,\dots,k}$ $(1 \le i \le 2^k)$ the 2^k different choices of ± 1 in $\{1,-1\}^k$ and define for $j=1,\dots,k$

$$f_j = x_o \sum_{i=1}^{2^k} \sigma_{ij}\, 1_{E_i}.$$

Then $\int M(|f_j|)d\mu = \sum \int_{E_i} M(x_o)d\mu = 0$ and therefore $\|f_j\|_M \le 1$ for each j. If $v_1,\dots,v_k$ is any sequence of ± 1, then there exists an i_o with $\sigma_{i_o j} = v_j$ $(1 \leq j \leq k)$ and so

$$\int M\left(\frac{|\sum v_j f_j|}{k(1-\varepsilon)}\right) d\mu = \sum \int_{E_i} M\left(\frac{x_o |\sum v_j \sigma_{ij}|}{k(1-\varepsilon)}\right) d\mu$$
$$\geq M\left(\frac{x_o}{1-\varepsilon}\right) \mu(E_{i_o}) \geq 1 .$$

Therefore, $\| \sum v_j f_j \|_M \geq k(1-\varepsilon)$ and varying ε and k gives a contradiction.

Now consider the other case, and let $\varepsilon > 0$ and $k \in \mathbb{N}$. We can find disjoint sets E_j $(j=1,\ldots,k)$ and $x_j > 0$ satisfying $M(x_j)\mu(E_j) \leq 1$ and $(1-\varepsilon)^{-1} m(0)\, x_j\, \mu(E_j) \geq 1$. Clearly $f_j = x_j 1_{E_j}$ has norm ≤ 1 and for any sign combination $v_1,\ldots,v_k$ we have

$$\int M\left(\frac{|\sum v_j f_j|}{k(1-\varepsilon)}\right) d\mu = \sum \int_{E_j} M\left(\frac{x_j}{k(1-\varepsilon)}\right) d\mu$$
$$\geq \sum \mu(E_j)\, m(0)\, \frac{x_j}{k(1-\varepsilon)} \geq 1 ,$$

showing that $\| \sum v_j f_j \|_M \geq k(1-\varepsilon)$.

It has been shown so far, that the given Orlicz function M always has a complementary Orlicz function which we denote by N. Considering the different cases we shall show now that both have the delta-property.

First let $\mu(\Omega) = \infty$. Assume that M does not satisfy the δ_2-condition and that $L_M(\mu)$ is (k,ε)-convex. Then it is immediate that for every $K > 0$ and $x_o > 0$ there exists an $0 < x \leq x_o$ such that

$$M((1-\varepsilon)^{-1}x) \geq K\, M(x).$$

By our assumptions on the measure space (case (c)) $0 < C :=$ $\liminf\{\mu(A) \mid A \text{ is an atom}\} < \infty$, if this set is non-empty; set $C = 2$ in the other case. Choose $x_n > 0$ satisfying $M((1-\varepsilon)^{-1}x_n) \geq 2^{n+1} M(x_n)$ and $M(x_n) \leq 2^{-n-1} C^{-1}$ $(n=1,2,\ldots)$. Next we can find disjoint sets E_n with $M(x_n)\,\mu(E_n) \leq 2^{-n}$ and with maximal measure respecting this condition, which means that also $M(x_n)\mu(E_n) \geq 2^{-n-1}$. Denote by $(\sigma_{in})_{1\leq i\leq k}$ the 2^k different sign combinations as before and define

$f_i = \sum_{n=1}^{2^k} \sigma_{in} x_n 1_{E_n}$ $(1 \leq i \leq k)$. Then

$$\int M(|f_i|)\, d\mu = \sum_{n=1}^{2^k} M(x_n)\mu(E_n) \leq \sum_{n=1}^{2^k} 2^{-n} \leq 1 ,$$

i.e. $\|f_i\|_M \leq 1$ $(1 \leq i \leq k)$, and for any sign combination $v_1,\ldots,v_k$

there exists n_o with $\sigma_{in_o} = v_i$, so that

$$\int M(\frac{|\sum v_i f_i|}{k(1-\varepsilon)})d\mu = \sum_{n=1}^{2^k} \int_{E_n} M(\frac{x_n|\sum v_i \sigma_{in}|}{k(1-\varepsilon)})d\mu$$

$$\geq \mu(E_{n_o})\, M(\frac{x_{n_o}}{1-\varepsilon}) \geq 2^{n_o+1}\, \mu(E_{n_o})\, M(x_{n_o}) \geq 1 \, .$$

Therefore $\| \sum v_i f_i \|_M \geq k(1-\varepsilon)$, which is a contradiction.

Assume now that M does not satisfy the Δ_2-condition, while μ has a nonatomic part or is finite. Let $\varepsilon > 0$ and $k \in \mathbb{N}$. Again, for every $K > 0$ and $x_o > 0$ there exists an $x > x_o$ such that $M((1-\varepsilon)^{-1}x) \geq K\, M(x)$. To apply this, choose x_n so that $M((1-\varepsilon)^{-1}x) \geq 2^{n+1}M(x)$ and $M(x_n) \geq p_n$ (where $\sum p_n^{-1} 2^{-n} \leq \mu(\Omega)$). Obviously there exist disjoint sets E_n satisfying $2^{-n-1} \leq \mu(E_n)M(x_n) \leq 2^{-n}$. Proceeding as above with the same notations and the same f_i's, one obtains

$$\int M(|f_i|) = \sum_n M(x_n)\mu(E_n) \leq 1 \, ,$$

i.e. $\|f_i\|_M \leq 1$, and

$$\int M(\frac{|\sum v_i f_i|}{k(1-\varepsilon)})d\mu \geq \mu(E_{n_o})\, M(\frac{x_{n_o}}{1-\varepsilon}) \geq 1 \, ,$$

i.e. $\| \sum v_i f_i \|_M \geq k(1-\varepsilon)$, a contradiction.

Since the proof for M is finished it remains to consider N. Let $H_M(\mu) := \{f \in L_M(\mu) \mid \int M(t|f|)d\mu < \infty$ for all $t > 0\}$.

It has been shown by Luxemburg that $H_M(\mu)$ is a closed subspace of $L_M(\mu)$ ([11], Lemma 1, p.55), and that $H_M(\mu)^*$ is isomorphic to $L_N(\mu)$ ([11], Theorems 1 and 2, p.55/56). Therefore it follows from Gisey ([5]), that $L_N(\mu)$ is B-convex, since $H_M(\mu)$ is. What has been shown above for M, proves the delta-property for N.

q.e.d.

<u>Remarks</u>: Since Theorem 1 is contained in Theorem 2, the classification of B-convex Orlicz spaces can be derived using results of Akomovic, Beck, Gisey and Luxemburg, that is, no deeper subspace theory is needed.

There are some more equivalent statements to those of the theorems which are well-known and of which we finally want to state a few important ones. It has been shown by Luxemburg ([11], Theorem 5, p.60) that for complementary Orlicz functions M and N, $L_M(\mu)$ is reflexive if and only if M and N have the delta-property. Hence the B-convex Orlicz spaces are exactly the reflexive ones. In [3] Enflo has proved that for a Banach space, superreflexive is equivalent to uniformly convexifiable (and this is equivalent to uniformly smootheable), hence super-reflexibility is also an equivalent condition. Some other super-properties are also equivalent: super RNP and super MCT (see Pisier [14] or [15]). Finally note that separability of $L_M(\mu)$ and $L_N(\mu)$ is only equivalent in the case of a separable measure μ (see Luxemburg [11], Theorem 6, p.61).

Literature

[1] V. AKIMOVIČ: On uniformly convex and uniformly smooth Orlicz spaces. Teoria Funkciǐ Funk.Anal. i Pril. 15 (1970), 114-120 (in Russian).

[2] A.BECK: On the strong law of large numbers. Ergodic Theory, Academic Press, New York 1963.

[3] P.ENFLO: Banach spaces which can be given an equivalent uniformly convex norm. Israel J.Math. 13 (1972), 281-288.

[4] Y.GRIBANOV: On the theory of l_M-spaces. Uc.zap.Kazansk. un-ta, 117 (1957), 62-65 (in Russian)

[5] D.P.GISEY: On a convexity condition in normed linear spaces. Trans.Amer.Math.Soc. 125 (1966), 114-146.

[6] M.A.KRASNOSELSKII, Y.B.RUTICKII: Convex functions and Orlicz spaces. Groningen, Netherlands (1961), translated from Russian.

[7] K.J.LINDBERG: On subspaces of Orlicz sequence spaces. Studia Mathematica 45 (1973), 119-146.

[8] J.LINDENSTRAUSS,L.TZAFRIRI: Classical Banach spaces. Lecture Notes in Mathematics 338 (1973), Springer-Verlag.

[9] ---, --- : Classical Banach Spaces I. Sequence Spaces; Ergebnisse der Math. Vol.92, Springer (1977).

[10] ---, --- : On Orlicz sequence spaces I-III. Israel J.Math. 10 (1971), 379-390, 11 (1972), 355-379, 14 (1973), 368-389.

[11] W.A.J.LUXEMBURG: Banach function spaces. Thesis, Assen,Netherlands (1955).

[12] ---, A.C.ZAANEN: Riesz spaces I. Amsterdam 1971.

[13] H.W.MILNES: Convexity of Orlicz spaces. Pacific J.Math. 7 (1957), 1451-1483.

[14] G.PISIER: Martingales à valeur dans les espaces uniformement convexes.

[15] W.A.WOYCZYŃSKI: Geometry and martingales in Banach spaces. Probability-Winter School, Karpacz, Poland (1975), 229-275.

Lower layers in $\mathbb{R}^2$ and convex sets in $\mathbb{R}^3$ are not GB classes*

R. M. Dudley

Abstract. In $\mathbb{R}^2$ let $\langle u,v\rangle \le \langle x,y\rangle$ iff $u \le x$ and $v \le y$. A lower layer is a set $A \subset \mathbb{R}^2$ such that if $\langle u,v\rangle \le \langle x,y\rangle \in A$ then $\langle u,v\rangle \in A$. Let λ be Lebesgue measure on a bounded, open, non-empty set in $\mathbb{R}^2$. Let W be the Gaussian process indexed by Borel sets with $EW(A) = 0$ and $EW(A)W(B) = \lambda(A \cap B)$ (white noise).

It is proved that W has almost all sample functions unbounded on the collection LL of all lower layers, i.e. LL is not "GB." Likewise, in $\mathbb{R}^3$ the collection of all convex subsets of the unit ball is not GB.

1. Introduction. For any finite measure space (X, S, μ), let W_μ be the Gaussian process indexed by S with mean 0 and covariance

$$EW_\mu(A)W_\mu(B) = \mu(A \cap B).$$

Then W_μ has independent values on disjoint sets. A collection $A \subset S$ is called a GB class (for μ) iff W_μ has a version a.s. bounded on A, cf. [3]. If μ is a probability measure, the process defined by

$$G_\mu(A) := W_\mu(A) - \mu(A)W_\mu(X)$$

arises as a limit of normalized empirical measures, cf. [4]. Then G_μ has a version a.s. bounded on a class $A \subset S$ iff A is a GB class.

Here we will let $X = \mathbb{R}^2$ or $\mathbb{R}^3$, S = Borel sets.

*This research was partially supported by National Science Foundation Grant MCS76-07211 A01.

2. Lower layers. De Hardt (1970, 1971), in effect, and Steele (1978) considered the class $\mathcal{LL}$ of lower layers (note that if A is a lower layer, then $1_A(u,v) \geq 1_A(x,y)$ for $\langle u,v\rangle \leq \langle x,y\rangle$).

Prop. 1. For Lebesgue measure λ on the unit square in $\mathbb{R}^2$, $\mathcal{LL}$ is not a GB class.

Proof. Let $W := W_\lambda$. Let $T := T_{01}$ be the right triangle on which $x \leq 1$ and $y \leq 1 \leq x + y$. Let C_{01} be the square where $\frac{1}{2} \leq x \leq 1$ and $\frac{1}{2} \leq y \leq 1$.

For $n = 1, 2, \ldots,$ and $j = 1, \ldots, 2^n$, let T_{nj} be the right triangle defined by $x + y \geq 1$, $(j-1)/2^n \leq x < j/2^n$, and $1 - j/2^n < y \leq 1 - (j-1)/2^n$. Let C_{nj} be the square filling the upper right corner of T_{nj}, on which $(2j-1)/2^{n+1} \leq x < j/2^n$ and $1 - (2j-1)/2^{n+1} \leq y < 1 - (j-1)/2^n$. Since the squares C_{nj} are disjoint for all n and j, the random variables $W(C_{nj})$ are independent. We have $\lambda(C_{nj}) = 1/4^{n+1}$ for all n and j.

Let D be the diagonal where $x + y = 1$, $0 \leq x < 1$, so $0 < y \leq 1$. Let L_{nj} be the line segment of D included in T_{nj}, where $(j-1)/2^n \leq x < j/2^n$. Each point $p = \langle x,y\rangle \in D$ belongs to $T_{nj(n,p)}$ for some unique $j(n,p)$.

For each $p \in D$ and $M < +\infty$, the events

$$E_{np} := \{W(C_{nj(n,p)}) > M/2^{n+1}\}$$

are independent for $n = 1, 2, \ldots,$ and have the same positive probability. Thus, almost surely, such an event occurs. Let $n(p) := n(p,\omega)$ be the least such n, defined and finite for almost all ω.

Since the events E_{np} are measurable jointly in p and ω, Fubini's theorem implies that almost surely, for almost all p (with respect to uniform Lebesgue measure on D), some E_{np} occurs, so

$n(p) < +\infty$.

Let $V := V_\omega$ be the union of all the triangles $T_p := T_{n(p),j(n(p),p)}$. Let $C_p := C_{n(p),j(n(p),p)}$. Let $A_\omega := \{\langle x,y\rangle : x + y \leq 1\} \cup V_\omega$ and $B_\omega := A_\omega \setminus \bigcup_{p\varepsilon D} C_p$. Then A_ω and B_ω are lower layers. Almost all $p \varepsilon D$ belong to some interval which is the hypotenuse of a triangle T_p with the square C_p in its upper right, having $2W(C_p) > M/2^{n(p)}$. Hence, $W(A_\omega) - W(B_\omega) \geq M/2$ almost surely. Thus $\max(|W(A_\omega)|, |W(B_\omega)|) \geq M/4$. Letting $M \to +\infty$ we see that lower layers do not form a GB class, Q.E.D.

3. Convex sets. In $\mathbb{R}^3$, let λ be Lebesgue volume measure on some bounded set including the unit ball $B: x^2 + y^2 + z^2 \leq 1$. Let $W := W_\lambda$. I claimed the following in [3, p. 87, Remark], but without adequate proof until now.

Prop. 2. In $\mathbb{R}^3$, the collection of all convex sets is not a GB class for λ.

Proof. Let S^2 be the unit sphere $x^2 + y^2 + z^2 = 1$. On S^2 we have a longitude coordinate θ, $0 \leq \theta < 2\pi$, and a latitude coordinate z, $-\pi/2 \leq z \leq \pi/2$, where $\langle x,y\rangle = (x^2+y^2)^{1/2}\langle\cos\theta, \sin\theta\rangle$ and $z = \sin\phi$.

Let a quad be a subset $Q(\alpha,\beta;a,b)$ of S^2 where $\alpha \leq \theta < \beta$ and $a \leq \phi < b$. The associated plate $P(\alpha,\beta;a,b)$ will be the subset of B outside the convex hull of $S^2\setminus Q(\alpha,\beta;a,b)$.

We begin with the quad $Q_{01} := Q(0,\pi/4;0,\pi/4)$. At the nth stage, we decompose each $Q_{nj} = Q(\alpha_{nj},\beta_{nj};a_{nj},b_{nj})$ into 2^{16} quads by decomposing each interval $[\alpha_{nj},\beta_{nj}[$ and $[a_{nj},b_{nj}[$ into 256 equal parts. Thus we obtain quads $Q_{n+1,i}$, $i = 1, 2, \ldots, 2^{16(n+1)}$. For each n and j,

$$\beta_{nj} - \alpha_{nj} = b_{nj} - a_{nj} := 4\theta_n := \pi/2^{8n+2}.$$

For each quad Q_{nj}, let $v_{nj} \in S^2$ be the vector with coordinates $\theta = (\alpha_{nj} + \beta_{nj})/2$, $\phi = (a_{nj} + b_{nj})/2$. Let L_{nj} be a plane orthogonal to v_{nj}, passing through $(\cos\theta_n)v_{nj}$. Let C_{nj} be the cap $\{u \in B: (u,v_{nj}) > \cos\theta_n\}$, cut from B by L_{nj}.

<u>Lemma 1</u>. The boundary of Q_{nj} is entirely outside C_{nj}.

<u>Proof</u>. The smallest great-circle distance from the center v_{nj} to any point on the upper or lower boundary of Q_{nj}, where $\phi = a_{nj}$ or b_{nj}, is easily seen to be $2\theta_n > \theta_n$. Let ρ be the shortest great-circle distance from v_{nj} to any point with longitude θ differing by $2\theta_n$. Then ρ equals half the great-circle distance from v_{nj} to a point at the same latitude ϕ and with longitude differing by $4\theta_n$, using symmetry. Thus $\rho = \sin^{-1}(\cos\phi\sin(2\theta_n))$. Since $0 \leq \phi \leq \pi/4$ and $0 \leq \theta < \pi/4$, we have $\sin^2\rho > \sin^2(\theta_n)$, so $\rho > \theta_n$, Q.E.D.

By Lemma 1, the plate P_{nj} includes the cap C_{nj}, so

$$\lambda(P_{nj}) \geq \lambda(C_{nj}) = \pi\left(\frac{2}{3} - \cos\theta_n + \frac{1}{3}\cos^3\theta_n\right).$$

For $\theta := \theta_n$, we have $|\theta| \leq 1$, so

$$1 - \theta^2/2 \leq \cos\theta \leq 1 - \theta^2/2 + \theta^4/24,$$

$$\lambda(C_{nj}) \geq \pi\left(\frac{2}{3} - (1 - \theta^2/2 + \theta^4/24) + (1 - \theta^2/2)^3/3\right)$$

$$= \pi(5\theta^4 - \theta^6)/24 \geq \theta_n^4/2.$$

For an upper bound of $\lambda(P_{nj})$, we take the cap

$$D_{nj} := \{u \in B: (u,v_n) > \cos\psi\}$$

where $\psi := 4\theta_n$. Then $D_{nj} \supset P_{nj}$, so

$$\lambda(P_{nj}) \le \lambda(D_{nj}) = \pi\left(\frac{2}{3} - \cos\psi + \frac{1}{3}\cos^3\psi\right)$$

$$\le \pi\left(\frac{2}{3} - 1 + \psi^2/2 + (1 - \psi^2/2 + \psi^4/24)^3/3\right)$$

$$\le \pi(\psi^4/8 + 3(\psi^2/2 - \psi^4/24)^2 - (\psi^2/2 - \psi^4/24)^3)/3.$$

Now $\psi := \psi_n = \pi/2^{8n+2} < 1$, so

$$\lambda(P_{nj}) < \pi(\psi^4/8 + 3\psi^4/4)/3 = 7\pi\psi^4/24 < \psi^4 = (4\theta_n)^4.$$

Let $S_{nj} := P_{nj} \setminus \bigcup_i P_{n+1,i}$. Note that $\theta_{n+1} = \theta_n/16^2$ and P_{nj} includes exactly 16^4 plates $P_{n+1,i}$, being disjoint from the others. Thus

$$\lambda(S_{nj}) \ge \theta_n^4(1/2 - 16^4(4/16^2)^4) > \theta_n^4/3.$$

Now each point p in the first quad Q_{01} belongs to quads $Q_{nj(p,n)}$, $n = 1, 2, \ldots$. Each quad Q_{nj} corresponds to an S_{nj}. The S_{nj} are all disjoint for distinct n or j. For each p, the events

$$E_{n,p} := \{W(S_{n,j(p,n)}) > M\lambda(S_{n,j(p,n)})^{1/2}\}$$

are independent and all have the same probability. These events are jointly measurable in p and ω. For each p, almost surely there is a least $n := n(p) = n(p,\omega) < +\infty$ such that $E_{n,p}$ holds. Thus, almost surely, $n(p) < +\infty$ for almost all p with respect to surface area measure on $Q_{01} \subset S^2$, by Fubini's theorem.

Let

$$U_\omega := B \setminus \bigcup_{p,n \ge n(p)} P_{nj(p,n)},$$

$$V_\omega := B \setminus \bigcup_{p,n > n(p)} P_{nj(p,n)},$$

where equivalently we could write S_{nj} in place of P_{nj}. Then U_ω and V_ω are convex for all ω.

The area of a quad Q_{nj} is $4\theta_n(\sin b_{nj} - \sin a_{nj}) < 16\theta_n^2$. Thus almost surely

$$W(U_\omega) - W(V_\omega) \geq MA/30$$

where $A = 2^{-5/2}\pi$, the area of Q_{01}. Hence

$$\min(|W(U_\omega)|, |W(V_\omega)|) \geq MA/60.$$

Letting $M \to \infty$ shows that the convex sets do not form a GB class, Q.E.D.

REFERENCES

1. DeHardt, J. (1970). A necessary condition for Glivenko-Cantelli convergence in E_n. Ann. Math. Statist. 41 pp. 2177-2178.
2. DeHardt, J. (1971). Generalizations of the Glivenko-Cantelli theorem. Ann. Math. Statist. 42 2050-2055.
3. Dudley, R. M. (1973). Sample functions of the Gaussian process, Ann. Probability 1 66-103.
4. Dudley, R. M. (1978). Central limit theorems for empirical measures (to appear in Ann. Probability).
5. Steele, J. Michael (1978). Empirical discrepancies and subadditive processes. Ann. Probability 6 118-127.

Room 2-245, M.I.T.
Cambridge, Mass. 02139, U.S.A.

INVARIANT MEASURES FOR LINEAR OPERATORS

by E. Flytzanis

Abstract: We consider measure preserving transformations defined by bounded linear operators T in Banach spaces. We relate properties of the two structures of T and for a special class of operators we characterize the m.p.t.'s they can define.

Introduction. We denote by B a separable complex Banach and by B^* its dual. If $T : B \to B$ is a bounded linear operator we say that T accepts an invariant m if there exists a probability measure m defined on the Borel σ-algebra of B for which $m(T^{-1}(\cdot)) = m(\cdot)$. For convenience we assume also that the support of m, consisting of the points whose all neighborhoods have nonzero measure, spans B. We say that m is of integrable norm if $\int_B ||x|| \, dm(x) < \infty$, where $||\cdot||$ denotes the norm function on B.

Recalling some notions from the theory of measure preserving transformations (m.p.t.) we have: If h is a m.p.t. in a probability space its eigenvalues are the complex numbers c for which the equation $f(h(\cdot)) = cf(\cdot)$ has non-trivial complex valued solutions. The eigenvalues of a m.p.t. form always a countable subgroup of the circle group and they coincide with the eigenvalues of the isometry $V : L_2 \to L_2$ defined by $Vf(\cdot) = f(h(\cdot))$. We say that h has complete point spectrum if L_2 is spanned by the eigenfunctions of V, or equivalently if a collection of eigenfunctions of h generates the σ-algebra of the space and in this case we also have that the corresponding collection of eigenvalues generates the group of eigenvalues of h [6, p.214]. In particular a m.p.t. having c.p.s. is necessarily invertible. A m.p.t. h is said to be ergodic if the only sets invariant under h are the trivial sets, i.e. $m(h^{-1}(A) \cap A) = 0$ implies $m(A) = 0$ or 1. Finally we mention that all statements with respect to m are understood to hold m-a.e.

In the present work we extend some of the results obtained in [3] using the same technique. In §1 we complete the characterization of the m.p.t.'s defined by the class of operators T having the property that a total set of functionals in B^* have bounded orbits under T. In §2 we study orbit structure properties for the general case of operators T accepting an invariant m of integrable norm. In §3 we answer a question raised in [3] concerning the norm integrability condition. We should note that the study of linear m.p,t,'s was initiated by the consideration of eigenoperators for m.p.t.'s [1].

§1. A class of operators. If B, B^* are as above, a set of functionals $\{x^*\} \in B^*$ is said to be total if $x^*(x) = 0$ for every $x^* \in \{x^*\}$ implies $x = 0$, or equivalently if $\{x^*\}$ spans B^* in its B-topology (weak * topology). Throughout this section we assume that $T : B \to B$ has the property that a total set of functionals has bounded orbits under T^*. The next lemma was obtained in [3] assuming norm integrability for the invariant measure.

Lemma 1 B,T are as above and m is a Borel probability measure invariant under T and with support spanning B. Then:

(i) The m.p.t. defined by T has c.p.s.

(ii) The eigenvectors of T^* span B^* in the B-topology and the eigenvalues of T^* are all of norm 1 and they generate the group of eigenvalues of the m.p.t. defined by T.

Proof: We denote by M the metric space of complex valued (measurable) functions on (B,m) equipped with the topology of convergence in measure as given by the usual metric $\rho(f,g) = \int_B (|f-g|/1+|f-g|)dm$. The linear map $K : B^* \to M$ defined by $Kx^* = x^*(\cdot)$ is injective because of the assumption that the support of m spans B. It is also continuous when restricted to bounded subsets of B^* equipped with the

B-topology. Indeed on such subsets the B-topology is metric because of the separability of B [2, p.426] and we can consider convergence of sequences. However convergence of a sequence in the B-topology of B^* means convergence everywhere on B which implies convergence in m-measure.

Let now $x^* \in B^*$ have bounded orbit under T^* and denote by $C(x^*)$ the closure of this obrit in the B-topology. Then $C(x^*)$ is compact and $K : C(x^*) \to M$ is injective continuous satisfying also the equation $KT^* = VK$ where $V : M \to M$ is the isometry defined by $Vf(\cdot) = f(h(\cdot))$. Denoting the image of $C(x^*)$ under K by S, we have that S is compact and invariant under the isometry V having also a dense orbit. It follows that $V : S \to S$ has the structure of a minimal rotation on a compact metric topological group [4]. It satisfies also the equation $X^*(V(\cdot)) = T^*X^*(\cdot)$ where $X^* : K^{-1} : S \to B^*$ is continuous in the B-topology. If μ is the Haar measure on S and $\{f_i\}$ the collection of characters, the weak * integrals $x_i^* = \int_S f_i(\cdot)X^*(\cdot)d\mu \in B^*$ are well defined [2, p.347]. It follows directly that the non zero among the functionals x_i^* are eigenvectors of T^* having eigenvalues of norm 1 and their span in the B-topology coincides with that of $C(x^*)$.

By the above and the assumption on the orbits of T^* it follows that B^* is spanned in its B-topology by the collection of functionals that are eigenvectors of T^*. Considering them as functions on (B,m) we note that they are eigenfunctions of the m.p.t. defined by T having the same eigenvalues which in particular must all be of norm 1. Also they generate the Borel σ-algebra of B because of the separability of B [7, p.74] and the theorem then follows by the remarks in the introduction. Q.E.D.

It is not known whether the necessary conditions expressed in part (ii) of the Lemma are also sufficient. The next theorem gives such necessary and sufficient conditions for the less general case where the invariant measure is of integrable norm, generalizing corresponding results for contractions [5]. We remark that the equivalence between parts (i) and (ii) was shown in [3].

Theorem 1. B is a complex separable Banach space and $T : B \to B$ a bounded linear operator with the property that a total set of functionals have bounded orbits under T^*. Then the following are equivalent.

(i) T accepts an invariant m of integrable norm whose support spans B.

(ii) B is spanned by eigenvectors of T having eigenvalues of norm 1.

(iii) A dense set of elements in B have conditionally compact and recurrent orbits ($T^{n_i}x \to x$) under T.

Proof: (i) ⇒ (ii). By Lemma 1 and the construction in Theorem 2 below, [3]. (ii) ⇒ (iii). Clear, Omitted.

(iii) ⇒ (i). Let x be an element of B whose orbit under T is conditionally compact and recurrent. Denoting by X the closure of the orbit we have that $T:X \to X$ is unto because of the recurrence assumption and therefore induces an isometry V in the Banach space C(X) of continuous complex valued functions with the sup norm, where $Vf(\cdot) = f(h(\cdot))$. The linear map $\Lambda : B^* \to C(X)$ defined by $\Lambda x^* = x^*(\cdot)$ is compact by Ascoli's theorem [2, p.266], and satisfies the equation $\Lambda T^* = V\Lambda$. It follows by the assumption on T^*, the compactness of Λ and the properties of weak convergence in C(X) [2, p.264, Cor. 4] that the image $\Lambda(B^*) \subset C(X)$ is spanned by functions whose orbits under the isometry V are conditionally compact, or equivalently by the eigenfunctions of V [5]. Noting that $\Lambda(B^*)$ separates the points of X, that products of eigenfunctions of V are also eigenfunctions and that the constant functions are also eigenfunctions of V it follows that C(X) is spanned by the eigenfunctions of V [2, p.274]. Therefore $T : X \to X$ has the structure of a minimal rotation on a compact metric topological group [4] and accepts an invariant m with support X. Taking now a countable bounded collection $\{X_i\}$ as above, spanning B, we obtain the corresponding invariant measures m_i and then the measure $m = \Sigma a_i m_i$, where $a_i > 0$, $\Sigma a_i = 1$ satisfies part (i) of the theorem. Q.E.D.

§2. Orbit structure properties of T,T^*. In this section we generalize the necessary conditions obtained above to the general case of a bounded linear operator $T : B \to B$ accepting an invariant m.

Theorem 2. If $T : B \to B$ is a bounded linear operator accepting an invariant m whose support spans B, then:

(i) The eigenvalues of T^* have norm 1 and the functionals that do not lie in the subspace of B^* spanned by the eigenvectors of T^* in the B-topology have unbounded orbits under T^*.

(ii) If in addition the invariant m is of integrable norm then T leaves invariant a compact convex set $X \subset B$ with the property $\liminf_{n\to+\infty} ||T^n x - T^n y|| = 0$ for every pair $x,y \in B$. Also X together with the eigenvectors of T having eigenvalues of norm 1 span B.

Proof: (i) By the proof of Lemma 1, Omitted.(ii). Let m be as in the theorem. We consider the space $L_\infty(B,m)$ and the isometry $V : L_\infty \to L_\infty$ defined by $Vf(\cdot) = f(h(\cdot))$. Using well known properties of the isometry V extended naturally to the Hilbert space $L_2(B,m)$ [9, p.40], we can show the following: If $\{\phi(\cdot)\}$ denotes the collection of eigenfunctions of V in L_∞ and

$$H = \{f : f \in L_\infty, \int_B f\cdot\phi \, dm = 0 \text{ for every } \phi \in \{\phi\}\}$$

then: 1. H is a closed subspace of L_∞ in the L_1-topology. 2. $H \cup \{\phi\}$ span L_∞ in the L_1-topology. 3. For each $f \in H$, $V^{n_i} f \to 0$ in the L_1-topology for some sequence $n_i \to +\infty$.

We need also the following construction from [3]. Consider the linear map $K : L_\infty \to B$ defined by the strong integral $Kf = \int_B f(x) x dm(x) \in B$. It is compact because of the norm integrability condition [8] so its adjoint $\Lambda : B^* \to L_1 \subset L_\infty^*$ given by $\Lambda x^* = x^*(\cdot)$ is also compact. Noting that K is itself the adjoint of Λ and given also that L is separable it follows that K sends bounded sequences converging in the L_1-topology of L_∞ into sequences converging in the strong topology of B [2. p.486].

It follows directly from the above that if S is the closed unit ball in H then its image $X = K(S)$ has all the required properties. Q.E.D.

The above theorem takes a simple form in the case where the m.p.t. defined by T is weakly mixing, i.e. ergodic without any eigenvalues except the trivial 1. We note first that if m, invariant under T, is of integrable norm then we can assume w.l.o.g. that the mean $\bar{x} = \int_B x\,dm$ is zero because $T\bar{x} = \bar{x}$ and we can consider the restriction of T to the subspace spanned by the translate of m by $\bar{x}$.

Corollary 1. If T accepts an invariant m of integrable norm and zero mean whose support spans B and such that the m.p.t. so defined by T is weakly mixing, then:

(i) $\limsup ||T^{*n}x^*|| = +\infty$ for every $x^* \in B^*$.

(ii) T leaves invariant a convex compact set $X \subset B$ spanning B and such that $\liminf ||T^n x - T^n y|| = 0$ for every pair $x, y \in X$.

Remark. If the m.p.t. defined by T in the above corollary is in fact strongly mixing in (i) we have $\lim ||T^{*n}x^*|| = +\infty$ and in (ii) we have $\lim ||T^n x - T^n y|| = 0$.

Given a m.p.t. defined on a probability space we remark that using a construction similar to the one in the first example of [3] we can always realize it by a bounded linear operator in an infinite dimensional Banach space. In view of the apparent importance of the integrability condition the question was raised in [3] as to whether there exist a class of ergodic m.p.t.'s on a nonatomic probability space such that any realization of them by bounded linear operators will result in the invariant m having integrable norm. In particular assuming that $T : B \to B$ accepts an invariant m with no atoms we note that the norm function $\chi(T(\cdot))/\chi(\cdot)$ is ess. bd.. Hence the question arises as to whether this condition implies that $\chi(\cdot)$ is integrable for some class of ergodic m.p.t.'s. The answer is negative as indicated by the construction below.

Example. Let T be an ergodic m.p.t. on a nonatomic probability space. We show the stronger statement that there exists a non-negative and non-integrable function $\chi(\cdot)$ for which $\chi(T(\cdot)) - \chi(\cdot)$ is ess. bd. By Rokhlin's theorem, valid also in the noninvertible case, for every integer q we can find a set A such that $A_k = h^{-k}A$, $k = 1, 2, \ldots, q$ are disjoint and $o < m(co(\bigcup_1^q A_i)) < 1/2$. Given q,A and choosing also an integer $p = 2r \leq q$ and a real $c > o$ we define a non-negative $f(\cdot)$ by $f(s) = kc$ if $s \in A_k \cup A_{p+1-k}$, $k = 1, 2, \ldots r$ and zero elsewhere. We have:

1. $f(T(\cdot) - f(\cdot) \leq c$, 2. $\int_S f(\cdot)dm \geq r\, c/4q$, 3. $m(\text{supp. } f) \leq 2r/q$.

For $n = 1, 2, \ldots,$ we choose q_n, r_n, c_n, f_n as above and set $f = \Sigma f_n$. Then:

1. $f(T(\cdot)) - f(\cdot) \leq \Sigma c_n$, 2. $\int_S f dm \geq \Sigma c_n r_n^2/4q_n$, 3. $m(\{s : f(s) = \infty\}) \leq \Sigma_{n=i} 2r_n/q_n$

Choosing e.g., $c_n = 1/n^2$, $r_n = n^3$, $q_n = n^5$ we have $\Sigma r_n/q_n < \infty$ so $f(\cdot)$ is well defined by 3. and $\Sigma c_n < \infty$, $\Sigma c_n r_n^2/q_n = \infty$ so $f(\cdot)$ has the required properties by 1.,2.

Given now an ergodic m.p.t. we construct $f(\cdot)$ as above and then using a construction similar to the one in the first example of [3] we can find an equivalent m.p.t. defined by a linear operator with respect to an invariant measure of non-integrable norm. Of course it may still be the case that the invariant m is of integrable norm if we use particular classes of operators, e.g. isometries. It is not known what the case is for the class of operators considered in §1.

Finally I should mention that some of the work above was done during the author's visit at the University of Warwick.

References

1. Beck A. Eigenoperators of ergodic transformations, Trans. A.M.S. 94 (1960), 118-129.
2. Dunford N & Schwartz J. T. *Linear Operators*, Part I., Interscience Publishers (1966).
3. Flytzanis E. Linear operators as measure preserving transformations, Ann. Sci. Univ. Clermont. No. 15 (1977), 63-75.
4. Halmos P. R. & Von Neumann T. Operator methods in classical mechanics II. Ann. Math. 43, pp.235-247, (42).
5. Jacobs K. Lectures in ergodic theory, Vol. I. Aarhus Un. 1962/63.
6. Krengel U. Weakly wandering vectors and weakly independent partitions, Trans. A.M.S. 164 (1972).
7. Kuo H. H. Gaussian measures in Banach spaces. Springer-Verlag L.N. 463.
8. Uhl J. J. The range of a vector measure, Proc. A.M.S. 23 (69).
9. Halmos R. P. Lectures on ergodic theory, Math. Soc. Japan, No. 3 (1956).

University of Thessaloniki,
Thessaloniki,
Greece.

October 1978

ON SUMS OF INDEPENDENT RANDOM VARIABLES WITH VALUES IN L_p $(2 \leq p < \infty)$

Evarist Giné[1)]
Instituto Venezolano de Investigationes Cientifices

V. Mandrekar[2)] and Joel Zinn[3)]
Michigan State University

0. Introduction. The purpose of this note is to study some limit theorems for sums of independent random variables taking values in $L^p = L^p(S,\mathcal{B},\nu)$, the space of equivalence classes of real functions f on a σ-finite measure space $(S,\mathcal{B},\mu)$ for which $|f|^p$ is integrable with $2 \leq p < \infty$. By a random variable we mean a strongly measurable function on a probability space (Ω,F,P) into L^p. As we shall be dealing with at most countably many random variables we shall assume without loss of generality (see [3], p. 168) that $L_p(S,\mathcal{B},\nu)$ is separable and all measurability concepts for L_p-valued functions coincide.

In [10], a complete solution to the Central Limit Theorem in L^p was given in the i.i.d. case. For the general Central Limit Problem involving triangular arrays of (not necessarily i.i.d.) L^p-valued random variables, some sufficient conditions were given in ([4], [7]).

We base our main results on the following theorem due to H.P. Rosenthal [11].

0.1 Theorem. Let $2 \leq p < \infty$. Then there is a constant $K_p < \infty$ so that for any sequence $\{X_j\}_{j=1}^n$ of independent real-valued random variables with $E|X_j|^p < \infty$ and $EX_j = 0$ $(j = 1,2,\ldots,n)$ we have for all $n \geq 1$,

$$\frac{1}{2}\max\{(\sum_{j=1}^n E|X_j|^p)^{\frac{1}{p}}, (\sum_{j=1}^n E|X_j|^2)^{\frac{1}{2}}\} \leq (E|\sum_{j=1}^n X_j|^p)^{\frac{1}{p}} \leq K_p \max\{(\sum_{j=1}^n E|X_j|^p)^{\frac{1}{p}}, (\sum_{j=1}^n X_j|^2)^{\frac{1}{2}}\}.$$

We first use Theorem 0.1 to derive an analogue of the "three series theorem" for L^p-valued random variables. We note that sufficient conditions for convergence of series in type 2 spaces were given by N. Jain [6]. Our "three series theorem"

---

1) This author's research was carried out at the Universitat Autonoma de Barcelona, partially supported by research funds there and CONICIT Grant 51-26.S1.0893.

2 Supported in part by NSF Grant No. MCS78-02878.

3) Supported in part by NSF Grant No. MCS77-21090.

uses four series. The next theorem we prove is the analogue of the general Weak Law of Large Numbers (WLLN). We also indicate how this result can be obtained from the criteria for the convergence of series. Finally, we obtain a solution of the general Central Limit Problem for triangular arrays from the weak law of large numbers. Thus, at least in this special case, we show how the necessary and sufficient conditions for the convergence of a series of independent Banach space valued random variables yield a complete solution to the Central Limit Problem. As applications, first we derive from our work on the Central Limit Problem, the CLT in [10] using the following.

0.2 Lemma ([10], p. 302). Let B be a real separable Banach space and X be a B-valued random variable. Then for all $n \geq 1$ and $\delta > 0$ and $2 < p < \infty$ we have

$$n E\|\frac{X}{\sqrt{n}} I(\|X\| \leq \delta\sqrt{n})\|^p \leq \frac{p}{p-2} \delta^{p-2} \Lambda^2(X),$$

where $\Lambda^2(X) = \sup_{t>0} t^2 P(\|X\| > t)$.

Proof. $E\|X\|^p I(\|X\| \leq \delta\sqrt{n}) \leq \int_0^{(\delta\sqrt{n})^p} P(\|X\| > t^{\frac{1}{p}}) dt$

$$\leq \int_0^{\delta^p n^{p/2}} \frac{\Lambda^2(X)}{t^{2/p}} dt = \frac{p}{p-2} \Lambda^2(X)(\delta\sqrt{n})^{p-2}.$$

As a second application, we derive a result of Yurinski ([12], Theorem 3) on the Levy-Khinchine representation of infinitely divisible law on L_p ($p \geq 2$). We note that this result in [12] is based on a result of Novikov, whose complete proof is not available. We present Novikov's result as a consequence of Theorem 0.1 using ideas used in this paper.

Throughout this paper we shall be dealing with measures on L_p. For the analysis involved we use some approximations in L^p and a characterization of compact sets in L^p. For this we need the following facts from [3]. Let $(S, \mathcal{B}, \nu)$ be a σ-finite measure space. For any subfamily $\pi = \{E_1, \ldots, E_n\} \subseteq \mathcal{B}$ of disjoint sets of finite ν-measure, U_π will denote the map from $L^p \to L^p$ given by

$$U_\pi(f) = \sum_1^n [\frac{1}{\nu(E_j)} \int_{E_j} f(s)\nu(ds)] I_{E_j} .$$

Then for each π, U_π is linear, has finite dimensional range and $U_\pi \uparrow I$ (the identity in L^p) where $\{\pi\}$ is directed by "refinement".

0.3 Lemma. A set $K \subseteq L^p$ is compact iff (a) K is bounded and (b) $U_\pi f \to f$ in L^p uniformly in $f \in K$. Further, if $L_p(S,\mathcal{B},\nu)$ is separable then (b) can be replaced by (b') $\exists$ a countable family $\{U_{\pi_n}\}_{n=1}^\infty$ such that $U_{\pi_n} f \to f$ in L^p uniformly in $f \in K$ and π_{n+1} is a refinement of π_n.

The next corollary is well-known.

0.4 Corollary. Let $(S,\mathcal{B},\nu)$ be a separable measure space and μ a σ-finite Borel measure on Borel subsets $\mathcal{B}(L_p)$. Then there exists a $\mu \otimes \nu$ measurable map $e: L_p \otimes S \to R$ such that $e(f,\cdot) \in f$. We denote $e(f,t) = \hat{f}(t)$.

The following lemma is an immediate consequence of Lemma 0.3 and Prohorov's Theorem.

0.5 Lemma. A bounded sequence of measures $\{G_n\}$ is relatively compact iff for all $\delta > 0$

(a) $\lim\limits_{r\to\infty} \sup\limits_n G_n\{\|x - U_{\pi_r}(x)\| > \delta\} = 0$

(b) $\lim\limits_{\delta\to\infty} \sup\limits_n G_n\{\|x\| > \delta\} = 0$.

We end this section by giving the following notation and terminology.

Let B denote a real seaprable Banach space. A triangular array of B-valued random variables will be denoted by $\{X_{nj}\}_{j=1}^{k_n}$ $(n = 1,2,\ldots)$, where $\{X_{nj}\}_{j=1}^{k_n}$ are independent for each n. A triangular array will be called uniformly infinitesimal (u.i.) if for every $\varepsilon > 0$, $\max\limits_{1\le j\le k} P(\|X_{nj}\| > \varepsilon) \to 0$. Here by B-valued random variable we mean a measurable function from a probability space (Ω,F,P) into B. Associated with X we get the symmetrization $\tilde{X}$ of X written as $\tilde{X} = X - X'$ where X and X' are independent identically distributed. Associated with a triangular array we have,

$$S_n = \sum_1^{k_n} X_{nj},\ X_{nj\delta} = X_{nj}(\|X_{nj}\| \le \delta),\ X_{nj}^{\delta} = X_{nj}(\|X_{nj}\| > \delta)$$

$$S_{n\delta} = \sum_{j=1}^{k_n} X_{nj\delta},\ b_{nj} = EX_{nj1},\ \overline{X}_{nj} = X_{nj} - b_{nj}$$

$$\overline{S}_n = \sum_{j=1}^{k_n} \overline{X}_{nj},\ \underline{X}_{nj} = X_{nj1} - b_{nj};\ F_n = \sum_{j=1}^{k_n} \mathcal{L}(X_{nj})$$

$$F_n^{(\delta)}(\cdot) = \sum_{j=1}^{k_n} P(\|X_{nj}\| > \delta,\cdot),\quad F_{[\varepsilon,n]} = F|_{\{x:\ \varepsilon\le\|x\|\le n\}}$$

and finally

$$F^{(\delta)} = F|_{[\delta,\infty)}.$$

1. The Four Series Theorem and the WLLN. Throughout $\{X_j\}$ will denote a sequence of independent B (or L^p)-valued r.v.'s. Now define φ: $B \to B$ by

$$\varphi(x) = \begin{cases} x & \text{if } \|x\| \le 1 \\ \frac{x}{\|x\|} & \text{if } \|x\| > 1 \end{cases}.$$

It is easily checked that

(1.1)

(a) $\|\varphi(x) - \varphi(y)\| \le 2\|x-y\|$, $x,y \in B$

(b) for $\lambda > 1$, $|\varphi(\lambda x)(t)| \ge |\varphi(x)(t)|$ if $x \in L^p(\nu)$ and $x(t) \in R'$.

We have the (now) simple

1.2 Lemma. $\{\sum_{j=1}^{n} X_j\}$ converges in B a.s. iff the following three conditions hold.

(a) $\sum P(\|X_j\| > 1) < \infty$

(b) $\sum_{j=1}^{n} E\varphi(X_j)$ converges in B

(c) $E\|\sum_{j=N}^{n} \varphi(X_j) - E\varphi(X_j)\|^p \to 0$ as $n \ge N \to \infty$ for some (or for all) p, $1 \le p < \infty$.

Proof ($\Rightarrow$). (a) follows by Borel-Cantelli as usual since $X_j \to 0$ a.s. Again as usual, $\sum_{j=1}^{n} \varphi(X_j)$ converges a.s. Also,

$$\|\sum_{j=N}^{n} E\varphi(X_j)\| \le E\|\sum_{j=N}^{n} \varphi(X_j)\|$$

and for all $0 < p < \infty$

$$E \max_{N\le j\le n} \|\varphi(X_j)\|^p = \int_0^1 P(\max_{N\le j\le n} \|\varphi(X_j)\|^p > t)dt \to 0$$

by the Lebesgue Dominated Convergence Theorem. Hence by Hoffman-Jørgensen's Theorem 3.1 ([5], see particularly the bottom of p. 164 and top of p. 165) (b) and (c) follow.

(⇐) We need only recall that convergence in probability implies a.s. convergence for a series of independent r.v.'s.

Remark. Lemma 2.2 can be stated more simply by replacing (b) and (c) by

$$\lim_{\underline{n<N}\to\infty} E\|\sum_{j=N}^{n} \varphi(X_j)\|^p = 0 \quad \text{or} \quad \lim_{\underline{n>N}\to\infty} E\|\sum_{j=N}^{n} X_j I(\|X_j\| \leq 1)\|^p = 0.$$

However the above formulation will be more convenient to us as it is in the real-valued case.

1.3 Theorem. The following are equivalent for a family $\{X_j\}_{j=1}^{\infty}$ of independent L_p-valued random variables for $2 \leq p < \infty$.

(i) $\sum X_j$ converges a.s. in L_p

(ii)
(a) $\sum_{j=1}^{\infty} P(\|X_j\| > 1) < \infty$,

(b) $\sum_{j=1}^{n} E\varphi(X_j)$ converges in L^p,

(c) $\sum_{j=1}^{\infty} E\|\varphi(X_j) - E\varphi(X_j)\|^p < \infty$,

and

(d) $\int[\sum_{j=1}^{\infty} E|\widehat{\varphi(X_j)}(t) - (\widehat{E\varphi(X_j)})(t)|^2]^{p/2}\nu(dt) < \infty$.

(iii)
(a) $\sum_{j=1}^{\infty} P(\|X_j\| > 1) < \infty$,

(b') $\sum_{j=1}^{n} EX_j I(\|X_j\| \leq 1)$ converges in L^p

(c') $\sum_{j=1}^{\infty} E\|X_{j1} - EX_{j1}\|^p < \infty$

(d') $\int[\sum_{j-1}^{\infty} E|\hat{X}_{j1}(t) - E\hat{X}_{j1}(t)|^2]^{p/2}\nu(dt) < \infty$.

Proof. By Lemma 1.2 it suffices to show that (1.3c,d) are equivalent to (1.2c). But

$$E\|\sum_{j=N}^{n}\varphi(X_j) - E\varphi(X_j)\|^p = E\int|\sum_{j=N}^{n}\varphi(X_j)(t) - (E\varphi(X_j))(t)|^p\nu(dt))$$

$$= \int E|\sum_{j=N}^{n}\widehat{\varphi(X_j)}(t) - E(\widehat{\varphi(X_j)}(t))|^p\nu(dt).$$

The result now follows from Theorem 0.1. The proof (i) ⇔ (iii) is similar.

2. WLLN. In this section we obtain nasc for the WLLN under some mild conditions. Further we indicate how these results could have been obtained from the nasc for the convergence of series.

2.1 Theorem. Let $\{X_{nj}\}$ be a u.i. triangular array. Then in order that there exists $y_n \in L^p$ such that $S_n - y_n \overset{P}{\to} 0$ it is necessary and sufficient that the following conditions hold, where $b_{nj} = E(X_{nj}I(\|X_{nj}\| \leq 1)$

(i) $\sum_{j=1}^{k_n} P(\|X_{nj}\| > 1) \to 0$

(ii) $\sum_{j=1}^{k_n} E\|X_{nj}I(\|X_{nj}\| \leq 1) - b_{nj}\|^p \to 0$

(iii) $\int [\sum_{j=1}^{k_n} E|\hat{X}_{nj}(t)I(\|X_{nj}\| \leq 1) - b_{nj}(t)|^2]^{p/2}\nu(dt) \to 0.$

Proof. If $S_n - y_n \overset{P}{\to} 0$, then $S_n - S'_n \overset{P}{\to} 0$. But,

$$\frac{\sum_{j=1}^{k_n} P(\|\tilde{X}_{nj}\| > 1)}{1 + \sum_{j=1}^{k_n} P(\|\tilde{X}_{nj}\| > 1)} \leq P(\max_{1 \leq j \leq k_n} \|\tilde{X}_{nj}\| > 1) \leq P(\max_{1 \leq k \leq k_n} \|\sum_{j=1}^{k} \tilde{X}_{nj}\| > \tfrac{1}{2})$$
$$\leq 2P(\|\tilde{S}_n\| > \tfrac{1}{2}) \to 0.$$

Further, we then have

$$\sum_{j=1}^{k_n} P(\|X_{nj}\| > 1) = \sum_{j=1}^{k_n} \frac{P(\|X_{nj}\| > 1, \|X'_{nj}\| \leq \frac{1}{2})}{P(\|X'_{nj}\| \leq \frac{1}{2})} \qquad (\tfrac{0}{0} = 0)$$

$$\leq \frac{\sum_{j=1}^{k_n} P(\|X_{nj}\| > \frac{1}{2})}{1 - \max_{1 \leq j \leq k_n} P(\|X_{nj}\| > \frac{1}{2})} \to 0.$$

Hence under the u.i. condition (i) is necessary. On the other hand, under (i) $S_n - y_n \overset{P}{\to} 0$ iff $\sum_{j=1}^{k_n} X_{nj}I(\|X_{nj}\| \leq 1) - y_n \overset{P}{\to} 0$, which implies $\sum_{j=1}^{k_n} [X_{nj}I(\|X_{nj}\| \leq 1) - X'_{nj}I(\|X'_{nj}\| \leq 1)] \overset{P}{\to} 0$. Finally, by an application of

Hoffmann-Jørgensen's inequality (as in Lemma 1.2), and using Theorem 0.1 as in Theorem 1.3, we see that (ii) and (iii) are equivalent to

$$\sum_{j=1}^{k_n} [X_{nj} I(\|X_{nj}\| \le 1) - X'_{nj} I(\|X'_{nj}\| \le 1)] \xrightarrow{P} 0.$$

2.2 Corollary. Let $\{X_{nj}\}$ be a symmetric triangular array. Then $S_n \xrightarrow{P} 0$ iff (2.1(i), (ii) and (iii)) hold (noting that $b_{nj} = 0$ for all n,j).

We now indicate how Corollary (2.2) and then Theorem (2.1) can be obtained as "consequences" of Theorem 1.3. Suppose now that $\{Y_j\}$ are independent and symmetric with $P(\|X_j\| > 0) > 0$. Put $C(X) = \{a \in R : \sum_{j=1}^{n} a_j X_j$ converges a.s.$\}$. By Theorem 1.3 $a \in C(X)$ iff

$$(2.3)\quad \begin{cases} \text{(a)}\quad \Psi_1(a) = \sum_{j=1}^{\infty} E\|\varphi(a_j X_j)\|^p < \infty, \quad \text{and} \\ \text{(b)}\quad \Psi_2(a) = \int [\sum_{j=1}^{\infty} E|\widehat{\varphi(a_j X_j)}(t)|^2]^{p/2} \nu(dt) < \infty. \end{cases}$$

Hence (analogously to [13]) if we put

$$\Psi(a) = \Psi_1(a) + (\Psi_2(a))^{2/p} \qquad \text{and}$$

$$|a|_\Psi = \inf\{t > 0:\ \Psi(t^{-1}a) \le t\},$$

then $|\cdot|_\Psi$ is an F-norm and

$$|a^{(n)}| \to 0 \quad \text{iff} \quad \Psi(a^{(n)}) \to 0.$$

Then as in [5] the map

$$T:\ C(X) \to L_0$$

given by $T(a) = \sum_{j=1}^{\infty} a_j X_j$ is an isomorphism onto a closed subspace of L_0.

We may now obtain Corollary (2.2). First note that it suffices to consider all $\{X_{nj}\}$ as independent, then arranging $\{X_{nj}\}$ lexicographically as $\{Y_k\}$ and applying T to the $\{a^{(n)}\}$ given by

$$(a^{(n)})_j = \begin{cases} 1 & \text{if} \quad r_{n-1} < j \le r_n \\ 0 & \text{otherwise} \end{cases},$$

where $r_0 = 0$ and $r_n = \sum_{i=1}^{n} k_i$, we may use the continuity of T to obtain $|a^{(n)}|_\psi \to 0$ iff $T(a^{(n)}) = S_n \overset{P}{\to} 0$, which is essentially a restatement of Corollary (2.2).

3. CLT. In this section we obtain nasc conditions for a u.i. triangular array to satisfy the CLT, by using our results on the WLLN.

3.1 Theorem. Let $\{X_{nj}\}$ be a u.i. triangular array. Then in order that there exists $y_n \in L^p$ such that $\{S_n - y_n\}$ is relatively compact it is nas that the following conditions hold:

(i) For each $\delta > 0$, $\{F_n^{(\delta)}\}$ is relatively compact,

(ii) $\lim_{\ell\to\infty} \sup_n \sum_{j=1}^{k_n} E\|\hat{\underline{X}}_{nj} - U_{\pi_\ell}(\hat{\underline{X}}_{nj})\|^p = 0$

(iii) $\lim_{\ell\to\infty} \sup_n [\sum_{j=1}^{k_n} E|\hat{\underline{X}}_{nj}(t) - U_{\pi_\ell}(\hat{\underline{X}}_{nj})(t)|^2]^{p/2}\nu(dt) = 0$

and

(iv) $\{f(\overline{S}_n)\}$ is relatively compact for all $f \in L^{p*}$

in which case, $\{\sum_{j=1}^{k_n} (X_{nj} - E(X_{nj})I(\|X_{nj}\| \le 1))\}$ is relatively compact.

Proof. We first note that if a sequence $\{W_n\}$ is relatively compact, then $(I - U_{\pi_{\ell_n}})(W_n) \overset{P}{\to} 0$, where $\{\ell_n\}$ is any sequence $\uparrow \infty$. This follows from Theorem 5.5 of [2] and Lemma 0.3 (b'). Applying this to $\{S_n - y_n\}$ we obtain $(I - U_{\pi_{k_n}})(S_n - y_n) \overset{P}{\to} 0$. Hence by Theorem (2.1) for every $\delta > 0$

(v) $\sum_{j=1}^{k_n} P(\|X_{nj} - U_{\pi_{\ell_n}}(X_{nj})\| > \delta) \to 0$.

Also the first part of the proof of Theorem 2.1 shows that $\{F_n^{(\delta)}\}$ is a bounded sequence of measures. Hence by Lemma 0.5, $\{F_n^{(\delta)}\}$ is relatively compact for every $\delta > 0$.

The proofs of (ii) and (iii) are similar, so we demonstrate only (ii). If (ii) fails, there exist subsequences of $\{U_{\pi_\ell}\}$ and $\{k_n\}$, which we still write as $\{U_{\pi_n}\}$ and $\{k_n\}$, so that

(vi) $\sum_{j=1}^{k_n} E\|\hat{\underline{X}}_{nj} - U_{\pi_n}(\hat{\underline{X}}_{nj})\|^p > \eta \quad \forall\, n$.

However, if $\{S_n - y_n\}$ is relatively compact, we have by (i) (see [1]) that

$\{S_{n1} - y_n\}$ is relatively compact and hence $(I - U_{\pi_n})(S_{n1} - y_n) \overset{P}{\to} 0$. Now, since $\|\frac{1}{2}(I - U_{\pi_n})(X_{nj})I(\|X_{nj}\| \le 1)\| \le 1$, we may apply Theorem 2.1(ii)) which contradicts (3.1(vi)).

Again by (i) and [1] we have $\{\sum_{j=1}^{k_n} X_{nj}I(\|X_{nj}\| > \delta)\}$ is relatively compact for all $\delta > 0$. Hence we need only show that $\{\hat{S}_n\}$ is relatively compact. But by Lemma 0.5 we then need only show that for any $k_n \uparrow \infty$

(vii) $S_n - U_{\pi_{k_n}}(S_n) \overset{P}{\to} 0$.

But (3.1(v)) holds by (3.1(i)) and Lemma 0.5. Hence we may apply Theorem 2.1 to obtain (vii) completing the proof.

4. Applications. We now show how Theorem 3.1 can be used to obtain the central limit theorem given in Theorem 5.1 of [10] and the characterization of infinitely divisible laws (on L^p, ℓ^p) given by Yurinski in [12]. Finally, we remark on how Rosenthal's inequalities give a simple complete proof of a result of Novikov (see [12]).

4.1 Theorem ([10]). Let X be a mean zero r.v. with values in L^p $(2 < p < \infty)$ and $\{X_j\}$ i.i.d. copies of X. Then $\{S_n/\sqrt{n}\}$ converges in distribution iff

(i) $t^2 P(\|X\| > t) \to 0$

and

(ii) $\int [E|\hat{X}(t)|^2]^{p/2} \nu(dt) < \infty$.

Proof. The necessity of (i) has several simple proofs so we refer the reader to either of ([1], Theorem 2.10, [10], Theorem 3.3). Since (ii) is equivalent to "X is pre-Gaussian", it is trivially necessary.

For the sufficiency we first note that (4.1(i) and (ii)) imply that

(i') $t^2 P(\|\tilde{X}\| > t) \to 0$ as $t \to \infty$,

and

(ii') $\int [E|\hat{\tilde{X}}(t)|^2]^{p/2} \nu(dt) < \infty$.

To prove (3.1(ii)) we first note that for any $A > 0$

$$\Lambda^2(\tilde{X} - U_{\pi_n}(\tilde{X})) \le \varepsilon^2 + A^2 P(\|\tilde{X} - U_{\pi_n}(\tilde{X})\| > \varepsilon) + \sup_{t>A} t^2 P(\|2\tilde{X}\| > t).$$

Hence, $\Lambda(\tilde{X} - U_{\pi_n}(\tilde{X})) \to 0$ as $n \to \infty$. By Lemma 0.2, we then have

$$\lim_{\ell\to\infty} \sup_n n\, E\|\frac{\tilde{X}}{\sqrt{n}} I(\|\frac{\tilde{X}}{\sqrt{n}}\| \le 1) - E(\frac{\tilde{X}}{\sqrt{n}} I(\|\frac{\tilde{X}}{\sqrt{n}}\| \le 1))\|^p = 0,$$

which implies (3.1(ii)). To prove (3.1(iii)) we need only show that $\lim_{\ell\to\infty} \int [E|(I - U_{\pi_\ell})(\hat{X}(t))|^2]^{p/2} \nu(dt) = 0$. As this is demonstrated in the next proof in slightly greater generality we refer the reader to that proof. Now Theorem 3.1 shows that $S_n/\sqrt{n}$ centered is relatively compact. But the centerings are

$$n\, E(\frac{X}{\sqrt{n}} I(\|X\| \le \sqrt{n}))$$

which equals (since $EX = 0$) $-\sqrt{n}\, EXI(\|X\| > \sqrt{n})$. But

$$\sqrt{n}\, E\|X\|I(\|X\| > \sqrt{n}) = \sqrt{n} \int_0^\infty P(\|X\|I(\|X\| > \sqrt{n}) > t)dt \le \sqrt{n}\, P(\|X\| > \sqrt{n})$$

$$+ \int_{\sqrt{n}}^\infty P(\|X\| > t)dt \to 0$$

by (4.1(i)), so we're done.

4.2 <u>Theorem</u>. Let F be a non-negative measure on $\mathcal{B}(L^p)$ with $F(\{0\}) = 0$ and let

$$\varphi_F(x^*) = \exp \int [e^{ix^*(y)} - 1 - ix^*(y)I(\|y\| \le 1)]F(dy),\ x^* \in (L^p)^* .$$

Then φ is the characteristic functional of a probability measure μ on L^p iff

(i) $\int \|y\|^p \wedge 1\ F(dy) < \infty$

and

(ii) $\int [\int_{\|y\|\le 1} |\hat{y}(t)|^2 F(dy)]^{p/2} \nu(dt) < \infty$.

<u>Proof</u>. Since for any finite measure F, φ is always the characteristic functional of a probability measure, we may assume $F(\|X\| > 1) = 0$ and $F(L^p) = \infty$. Now choose $0 < \delta_n \downarrow 0$ such that $F(x:\ \delta_n < \|x\| \le 1) = k_n + t_n$, $0 \le t_n < 1$ and $k_n \uparrow \infty$. Then let $\{X_{nj}:\ 1 \le j \le k_n + 1\}$ be independent random variables with distribution $\{\mu_n\}$ with $\mu_n = \frac{1}{k_n + t_n} F|_{(\delta_n, 1]}$ for $1 \le j \le k_n$ and $\mu_{nk_n+1} = t_n/t_n+k_n\ F|_{(\delta_n,1]} + (1 - t_n)\delta_0$.

We now use Theorem 3.1 to show that $\{\sum_{j=1}^{k_n+1} \bar{X}_{nj}\}$ is relatively compact. Trivially (3.1(i)) holds and (3.1(ii) and (iii)) can be written as:

$$(4.3)\begin{cases}\text{(i)} & \lim\limits_{\ell\to\infty}\sup\limits_{n}\ [\int_{\delta_n<\|x\|}\|(I - U_{\pi_\ell})(x - \lambda_n)\|^p F(dx) = 0 \\ \text{(ii)} & \lim\limits_{\ell\to\infty}\sup\limits_{n}\ [\int_{\delta_n<\|x\|}|(I - U_{\pi_\ell})(\hat{x} - \lambda_n)(t)|^2 F(dx)]^{p/2}\nu(dt) = 0,\end{cases}$$

where $\lambda_n = \frac{1}{k_n + t_n}\int_{\delta_n<\|x\|} x\, F(dx)$.

But, letting $Q_\ell = I - U_{\pi_\ell}$, we have

$$\int_{\delta_n<\|x\|}\|Q_\ell(x - \lambda_n)\|^p F(dx) = \int_{\delta_n<\|x\|}\|Q_\ell(x) - \int_{\delta_n<\|z\|}Q_\ell(y)\frac{F(dy)}{k_n + t_n}\|^p F(dx)$$

$$\leq C_p[\int_{\delta_n<\|x\|}\|Q_\ell(x)\|^p F(dx) + \int_{\delta_n<\|z\|}\|Q_\ell(y)\|^p F(dy)]$$

with a similar inequality for (4.3(ii)).

Now $\|U_\pi(y)\| \leq \|y\|$ and if $\pi = \{A_1,\ldots,A_r\}$ and if $t \in A_j$

$$(\int|U_\pi(y)(t)|^2F(dy))^{\frac{1}{2}} = (\int|\int_{A_j} y(s)\frac{\nu(ds)}{\nu(A_j)}|^2F(dy))^{\frac{1}{2}}$$

$$= [\int\int_{A_j}\int_{A_j}\hat{y}(s)\hat{y}(u)\frac{\nu(ds)}{\nu(A_j)}\frac{\nu(du)}{\nu(A_j)}F(dy)]^{\frac{1}{2}}$$

$$\leq [\int_{A_j}\int_{A_j}(\int\hat{y}^2(s)F(dy))^{\frac{1}{2}}(\int\hat{y}^2(u)F(dy))^{\frac{1}{2}}\frac{\nu(ds)}{\nu(A_j)}\frac{\nu(du)}{\nu(A_j)}]^{\frac{1}{2}}$$

$$= \int_{A_j}(\int\hat{y}^2(s)F(dy))^{\frac{1}{2}}\frac{\nu(ds)}{\nu(A_j)} = U_\pi((\int\hat{y}^2(t)F(dy))^{\frac{1}{2}}).$$

But by Lemma 0.3 and 4.2(i), $U_{\pi_\ell}((\int\hat{y}^2(\cdot)F(dy))^{\frac{1}{2}})$ converges to $(\int\hat{y}^2(\cdot)F(dy))^{\frac{1}{2}}$ in L^p. Hence we have the uniform integrability of order p of $\{U_{\pi_\ell}((\int\hat{y}^2(\cdot)F(dy))^{\frac{1}{2}})\}_\ell$ and hence of

$$\{(\int|U_{\pi_\ell}(\hat{y})(\cdot)|^2F(dy))^{\frac{1}{2}}\}_\ell.$$

Further, for $t \in A_k$, $|U_\pi(y)(t)|^2 \leq \frac{1}{\nu(A_k)}\int|\hat{y}(s)|^2\nu(ds)$, implies $\int|\hat{y}(t) - U_{\pi_\ell}(y)(t)|^2F(dy) \to 0$ for ν - aa t.

Hence by the dominated convergence theorem and uniform integrability we have

$$\lim_{\ell\to\infty}\int\|y - U_{\pi_\ell}(y)\|^p F(dy) = 0$$

and

$$\lim_{\ell\to\infty}\int[\int|\hat{y}(t) - U_{\pi_\ell}(\hat{y})(t)|^2F(dy))^{p/2}\nu(dt) = 0.$$

Hence by Theorem 3.1 we have $\{\sum_{j=1}^{k_n+1} \bar{X}_{nj}\}$ is relatively compact. By applying linear functionals we see that φ_F is the limiting characteristic functional.

Conversely, if $\varphi_F = \hat{\mu}$, then the symmetrization, $\tilde{\mu}$, of μ has characteristic functional $\hat{\tilde{\mu}}(x^*) = \exp \int (\cos x^*(y) - 1)G(dy)$, where $G(A) = F(A) + F(-A)$. Further,

$$\int \|y\|^p G(dx) = 2\int \|y\|^p F(dx)$$

and

$$\int [\int |\hat{y}(t)|^2 G(d\hat{y})]^{p/2} \nu(dt) = 2^{p/2} \int [\int |\hat{y}(t)|^2 F(dy)]^{p/2} \nu(dt).$$

Hence it suffices to assume F and μ are symmetric. But then there exists a symmetric triangular array $\{X_{nj};\ 1 \le j \le n,\ 1 \le n < \infty\}$ such that for each n the X_{nj} have the same distribution and the distribution of S_n is μ. But then $nP(\|X_{n1}\| > t) \to F(\|x\| > t)$ for every t at which $F(\|x\| > \cdot)$ is continuous. Hence, (recalling $F(\|x\| > 1) = 0$),

$$\int \|x\|^p F(dx) = \int_0^1 F(\|x\|^p > t)dt$$

$$= \int_0^1 \underline{\lim}_n\, nP(\|X_{n1}\|^p > t)dt \le \underline{\lim}_n\, n \int_0^1 P(\|X_{n1}\|^p > t)dt$$

$$\le \sup_n n E\|X_{n1}\|^p = \sup_n \sum_{j=1}^n E\|X_{nj}\|^p$$

which is finite by Theorem 3.1 since $\{S_n\}$ is relatively compact. Hence we have (i). Similarly, condition (ii) is satisfied.

We now give a proof of Novikov's result. Here $\{N_j\}$ are i.i.d. Poisson with parameter 1.

4.3 Theorem. Assume $2 \le p < \infty$. Then there exists c_p and C_p such that for every real random variable, ξ, with characteristic function

$$\varphi(t) = \exp\{\int (\cos tu - 1)F(du)\}, \qquad (F \text{ symmetric})$$

$$\int |u|^p F(du) < \infty$$

and

$$\int |u|^2 F(du) < \infty$$

we have

$$c_p[\int |u|^p F(du) + (\int |u|^2 F(du))^{p/2}] \le E|\xi|^p \le C_p[\int |u|^p F(du) + (\int |u|^2 F(du))^{p/2}].$$

Proof. As in the proof of Theorem 4.2 we choose $0 < \delta_n \downarrow 0$ and $\{X_{nj}\}_{j=1}^{k_n+1}$

(real-valued). Letting $\{X^i_{nj}\}_{j=1}^{\infty}$ be i.i.d. copies of $\{X_{nj}\}$, we have $T_n = \sum_{j=1}^{k_n+1} \sum_{i=1}^{N_j} X^i_{nj}$ with $\{N_j\}$ independent of $\{X^i_{nj}\}$ has characteristic function

$$\varphi_n(t) = \exp(\int_{\delta_n<|x|} (\cos tx - 1)F(dx)).$$

Now T_n is a factor of ξ and hence by symmetry and Lévy's inequality

(4.3(i)) $$P(|T_n| > u) \leq 2P(|\xi| > u).$$

Then by first conditioning on $\{N_j\}$ and then applying Lemma 0.1 we have $E|T_n|^p$ bounded above (and below) by C_p (c_p) times $[\sum_{j=1}^{k_n+1} E|X_{nj}|^p + (\sum_{j=1}^{k_n+1} E|X_{nj}|^2)^{p/2}]$ which equals $[\int_{\delta_n<|x|} |x|^p F(dx) + (\int_{\delta_n<|x|} |x|^2 F(dx))^{p/2}]$. Now by writing $E|T_n|^p = \int_0^\infty P(|T_n|^p > u)du$ and using 4.3(i) we have the result by the dominated convergence theorem.

References

1. de Acosta, A., Araujo, A. and Giné, E. On Poisson measures, Gaussian measures and central limit theorem in Banach spaces. Advances in Prob. (Ed. P. Ney) (to appear).

2. Billingsley, P. Convergence of Probability Measures. John Wiley & Sons, 1968.

3. Dunford, N. and Schwartz, J.T. Linear operators I; General theory. Interscience, New York, 1958.

4. Hamedani, G.G. and Mandrekar, V. Central limit problem on L_p $(p \geq 2)$ II. Compactness of infinitely divisible laws. J. Multivariate Analysis 7 (1977) 363-373.

5. Hoffman-Jørgenson, J. Sums of independent Banach space valued random variables. Studia Math. 52 159-186.

6. Jain, N. Central limit theorem in a Banach space. Lecture Notes #526, 114-130, Springer-Verlag, New York, 1976.

7. Kuelbs, J. and Mandrekar, V. Harmonic analysis on F-spaces with a basis. Trans. Amer. Math. Soc. 169 113-152.

8. Loève, M. Probability Theory (Second Edition). D. Van Nostrand, New York, 1960.

9. Mandrekar, V. and Zinn, Joel. Central limit problem for symmetric case; Convergence to non-Gaussian laws. Studia Math. 67 (to appear).

10. Pisier, G. and Zinn, Joel. On the limit theorems for random variables with values in spaces L_p $(2 \leq p < \infty)$. Z. Wahrscheinlichkeitstheorie 41 (1978) 289-304.

11. Rosenthal, H.P. On the span in L^p of sequences of independent random variables. Sixth Berkeley Symposium on Math. Stat. and Prob. II, University of California Press, Berkeley and Los Angeles (1972) 149-167.

12. Yurinskii, V.V. On infinitely divisible distributions. Theor. Probability and Appl. 19 297-308.

13. Zinn, J. Another approach to the weak law of large numbers. Unpublished manuscript.

IVIC, Matematicas
Apartado 1827, Caracas, Venezuela

Department of Statistics and Probability
Michigan State University
East Lansing, Mihcigan 48824

The Generalized Domain of Attraction of a Gaussian Law on Hilbert Space[1]

Marjorie G. Hahn
Tufts University, Medford, MA 02163/USA

1. Introduction. The central limit problem for i.i.d. k-dimensional random vectors not necessarily assuming finite second moments has been shown in Hahn and Klass (1978) to require the use of norming matrices if a Gaussian limit law not concentrated on a lower dimensional hyperplane is desired. The object of this paper is to investigate the same problem for an infinite-dimensional real Hilbert space. The infinite-dimensional situation will constantly be compared to the Euclidean case in both its similarities and differences, a number of which are rather unexpected.

Throughout the paper, $X_1, X_2, \ldots$ denotes a sequence of independent mean zero Hilbert space-valued random vectors with the same law, $\mathcal{L}(X)$. S_n specifies the nth partial sum of the sequence. The law of X is called nondegenerate if for each unit vector θ, $\mathcal{L}(\langle X, \theta \rangle)$ is not concentrated at a single point.

Definition. X, or $\mathcal{L}(X)$, is said to be in the generalized domain of attraction (GDOA) of a Gaussian law, $\mathcal{L}(W)$, on $\mathcal{H}$ if there exists a sequence of linear operators T_n and elements $b_n \in \mathcal{H}$ such that

$$\mathcal{L}(T_n S_n + b_n) \to \mathcal{L}(W) .$$

Notice that when T_n is a constant multiple of the identity the GDOA reduces to the usual concept of domain of attraction (DOA) of a Gaussian. Since the central limit theorem is used for approximating joint distributions, considerable information is lost if the limit law has a degenerate component. Hence, we consider only nondegenerate Gaussian limit laws.

Section 2 is concerned solely with the Euclidean case. It contains a summary of the results of Hahn and Klass (1978) which completely characterize all laws in the GDOA of a nondegenerate Gaussian on $\mathbb{R}^k$. Furthermore, the moment structure of such laws is determined. Billingsley's convergence of types theorem for $\mathbb{R}^k$ is employed to characterize all suitable norming sequences of linear operators.

Section 3 is primarily a collection of examples indicating the disparity between the infinite and finite-dimensional situations. For example, X need not even possess a finite first moment in order to be in the GDOA of a nondegenerate Gaussian. Furthermore, the same sequence of partial sums may have one norming sequence of bounded linear operators and another norming sequence of unbounded linear operators.

The final section contains our main sufficiency theorem in the infinite-dimensional case. This theorem has two conditions, the first of which is shown to be necessary when the norming operators take a particular form.

2. Euclidean Case. A random variable X is in the DOA of the standard normal if there exist constants a_n for which $\mathcal{L}(S_n/a_n) \to N(0,1)$. This happens if and only if

(1) $$\lim_{t\to\infty} \frac{t^2 P(\|X\| > t)}{E(\|X\|^2 \wedge t^2)} = 0$$

in which case the constants are characterized by the relation

(2) $$\lim_{n\to\infty} n\,E(\|X\|^2 \wedge a_n^2)/a_n^2 = 1 .$$

It is easy to verify that in any Hilbert space (1) is equivalent to

(1') $M_X(t) \equiv E(\|X\|^2 \wedge t^2)$ is a slowly varying function of t, i.e. $\forall a \neq 0$, $\lim_{t\to\infty} M(at)/M(t) = 1$.

For nondecreasing functions, such as $M_X(t)$, it suffices to check the condition for $a = 2$.

Now, (1') and (2) together imply that the norming constants for a random variable in the DOA are of the form $a_n = \sqrt{n}\,h(n)$ where $h(n)$ is a slowly varying function of n .

In higher dimensions, normalization by constants or even componentwise often yields a degenerate limit (see examples in Hahn and Klass (1978)), hence the need for norming by linear operators. Since the limit must be nondegenerate, the linear operators will eventually be nonsingular.

It is clear that (1) is too weak a condition to imply that a k-dimensional random vector X is in the GDOA . For example, (1) can be satisfied when one coordinate of X is not even in the DOA . Just let X have independent coordinates (U,V) with $U \in$ DOA , $V \notin$ DOA and $\lim_{t\to\infty} M_V(t)/M_U(t) = 0$. X satisfies (1'), and hence (1), since

$$1 \le \lim_{t\to\infty} \frac{M_X(2t)}{M_X(t)} \le \lim_{t\to\infty} \frac{M_U(2t)+M_V(2t)}{M_U(t)}$$

$$= 1 + \lim_{t\to\infty} \frac{M_U(2t)}{M_U(t)} \frac{M_V(2t)}{M_U(2t)} = 1 .$$

The slightly stronger condition

$$\lim_{t\to\infty} \frac{t^2 P(|\langle X,\theta\rangle| > t)}{E(|\langle X,\theta\rangle| \wedge t^2)} = 0 \qquad \text{for all } \|\theta\| = 1$$

is also too weak because of the lack of control of the rate of convergence along the different directions (see Example 4, Hahn and Klass (1978)). Consequently, the appropriate solution must involve some uniformity.

Theorem 1. (Hahn and Klass (1978)). Let X be a mean zero, nondegenerate k-dimensional random vector. $X \in$ GDOA of the standard multivariate normal iff

$$(3) \qquad \lim_{t\to\infty} \sup_{\|\theta\|=1} \frac{t^2 P(|\langle X,\theta\rangle| > t)}{E(|\langle X,\theta\rangle|^2 \wedge t^2)} = 0 .$$

Moreover, the number $\sigma = \sup_{\|\theta\|=1} 1/P(\langle X,\theta\rangle \ne 0)$ is finite; and for $n \ge \sigma$ the following implicit definition uniquely specifies norming constants $a_n(\theta)$ for $\langle X, \theta \rangle$:

$$(4) \qquad a_n^2(\theta) = n\,E\,(|\langle X, \theta\rangle|^2 \wedge a_n^2(\theta)) \;.$$

There exist unit vectors $\gamma_{n,j}$ such that

$$a_n(\gamma_{n,1}) = \inf_{\|\theta\|=1} a_n(\theta)$$

$$a_n(\gamma_{n,j}) = \inf_{\|\theta\|=1,\, \theta \in \Gamma_{j-1}^{\perp}} a_n(\theta) \,, \quad j = 2,3,\ldots,k$$

where $\Gamma_{j-1} = \operatorname{span}\{\gamma_{n,1},\ldots,\gamma_{n,j-1}\}$. The norming linear operators T_n may be chosen to take the form

$$(5) \qquad T_n x = \sum_{j=1}^{k} (\langle x, \gamma_{n,j}\rangle / a_n(\gamma_{n,j}))e_j$$

where $\{e_j\}$ is the standard orthonormal basis.

The underlying idea of the proof and the construction of the specific operators T_n is that a degenerate limit will arise if one direction is normalized by constants which are too large. To avoid this, at stage n , a preferred orthonormal basis is constructed along which S_n will be normalized componentwise. The construction of these orthonormal bases is the reason for canonically defining the norming constants. Definition (4) gives rise to norming constants with the nice property that for $n \geq \sigma$, $\theta \mapsto a_n(\theta)$ is continuous. Consequently, there is a unit direction $\gamma_{n,1}$ in which $a_n(\theta)$ is minimal. $\gamma_{n,1}$ becomes the first basis element and successive basis elements $\gamma_{n,j}$ are determined by selecting a minimal limit direction for the $a_n(\theta)$ in the hyperplane perpendicular to the previously assigned basis elements.

With these orthonormal bases, for each n and $i > j$, the random variables $\langle X, \gamma_{n,i}\rangle$ and $\langle X, \gamma_{n,j}\rangle I_{(|\langle X, \gamma_{n,j}\rangle| \leq a_n(\gamma_{n,j}))}$ turn out to be uncorrelated. Furthermore, as a consequence of (3) and the continuity of $a_n(\theta)$, $\forall \varepsilon > 0$,

$$(6)\quad \lim_{n\to\infty} \sup_{\|\theta\|=1} \frac{n}{a_n(\theta)} E|\langle X, \theta\rangle| I_{(|\langle X, \theta\rangle| > \varepsilon a_n(\theta))} = 0 .$$

These two facts together imply that the truncated correlations

$$(7)\quad \frac{n}{a_n(\gamma_{n,j})a_n(\gamma_{n,i})} E\langle X, \gamma_{n,j}\rangle\langle X, \gamma_{n,i}\rangle I_{(|\langle X, \gamma_{n,j}\rangle| \leq a_n(\gamma_{n,j}),\ |\langle X, \gamma_{n,i}\rangle| \leq a_n(\gamma_{n,i}))} \to 0$$

which must indeed happen in order for the limit distribution to have an identity covariance matrix.

Notice that since $\langle X, \theta\rangle = \sum_{j=1}^{k} \langle X, \gamma_{n,j}\rangle\langle \gamma_{n,j}, \theta\rangle$, reasoning analogous to that in Proposition 6 of Hahn and Klass shows that for each θ, $a_n^2(\theta)$ is asymptotic to $\sum_{j=1}^{k} a_n^2(\gamma_{n,j})\langle \gamma_{n,j}, \theta\rangle^2$. Thus, componentwise norming along the preferred orthonormal bases yields proper normalization of each direction to a nondegenerate limit.

From the above theorem we can conclude that

Proposition 2. $X \in$ GDOA implies $E\|X\|^p < \infty$, $\forall p < 2$.

<u>Proof</u>. First notice that (3) implies (1), hence $M_X(t)$ is slowly varying, because

$$\sup_{\|\theta\|=1} t^2 P(|\langle X,\theta\rangle| > t)/E(|\langle X,\theta\rangle|^2 \wedge t^2)$$

$$\geq \max_{1\leq j\leq k} t^2 P(|\langle X,e_j\rangle| > t)/E(\|X\|^2 \wedge t^2)$$

$$\geq k^{-1}\, t^2 \sum_{j=1}^{k} P(|\langle X,e_j\rangle| > t)/E(\|X\|^2 \wedge t^2)$$

$$\geq k^{-1}\, t^2\, P(\|X\| > \sqrt{k}t)/E(\|X\|^2 \wedge t^2)\ .$$

Fix $0 \leq p < 2$. Let $0 < \delta < 2^{2-p} - 1$ and $\gamma = 1 + \delta$. There exists $t_o \geq 1$ such that $t \geq t_o$ implies

$$M_X(2t)/M_X(t) \leq \gamma \tag{8}$$

and

$$t^2 P(\|X\| > t)/M_X(t) \leq \delta\ . \tag{9}$$

Now, for $t \geq t_o$ and $p > 0$,

$$\begin{aligned} E\|X\|^p &= p\int_0^\infty u^{p-1}P(\|X\| > u)du \\ &= A + p\int_t^\infty u^{p-1}P(\|X\| > u)du\ , \quad A \text{ finite} \\ &= A + p\sum_{n=1}^{\infty}\int_{2^{n-1}t}^{2^n t} u^{p-1}P(\|X\| > u)du \\ &\leq A + \sum_{n=1}^{\infty}(2^n t)^p P(\|X\| > 2^{n-1}t) \end{aligned}$$

$$\leq \ A + \delta \sum_{n=1}^{\infty} 2^p(2^n t)^{p-2} M_X(2^{n-1}t) \quad \text{by (9)}$$

$$\leq \ A + \delta M_X(t) 2^p \sum_{n=1}^{\infty} (\gamma/2^{2-p})^{n-1} \quad \text{by (8)}$$

$$= \ A + \delta M_X(t) 2^p (1 - \gamma/2^{2-p})^{-1} < \infty \ . \qquad \square$$

Theorem 1 specifies one sequence of norming linear operators T_n .

Proposition 3. Any other sequence of norming linear operators A_n must be of the form $A_n = \delta_n \gamma_n T_n$, where γ_n are unitary operators and $\delta_n \to I$. Furthermore, $\lim_{n\to\infty} \|A_n\| = 0$.

Proof. The form of A_n follows from Theorem 4 of Billingsley (1966) and the fact that only the unitaries preserve the standard multivariate normal. Since $\|A_n\| \leq \|\delta_n\| \|T_n\|$ and $\lim_{n\to\infty} \|\delta_n\| = 1$, it suffices to show that $\lim_{n\to\infty} \|T_n\| = 0$. Notice that $T_n = D_n \cdot R_n$ where R_n is unitary and D_n is a diagonal matrix so $\|T_n\| = \|D_n\|$ = largest eigenvalue = $(a_n(\gamma_{n,1}))^{-1}$. Since $a_{n+1}(\gamma_{n+1,1}) > a_n(\gamma_{n+1,1}) \geq a_n(\gamma_{n,1}) > 0$ and $a_{2n}(\gamma_{2n,1}) > a_n(\gamma_{2n,1})$ we have

$$a_{2n}(\gamma_{2n,1})/a_n(\gamma_{n,1}) \geq a_{2n}(\gamma_{2n,1})/a_n(\gamma_{2n,1})$$

$$= \sqrt{2}\, M_X(a_{2n}(\gamma_{2n,1}))/M_X(a_n(\gamma_{2n,1}))$$

$$\geq \sqrt{2} \ .$$

Consequently, $\lim_{m\to\infty} a_m(\gamma_{m,1}) = \lim_{j\to\infty} a_{2^j n}(\gamma_{2^j n,1}) \geq$ $\lim_{j\to\infty} 2^{j/2} a_n(\gamma_{n,1}) = \infty$. Therefore, $\lim_{n\to\infty} \|T_n\| = 0$. □

Urbanik ((1972), Proposition 3.1) has derived the fact that $\lim_{n\to\infty} \|A_n\| = 0$ using abstract reasoning rather than specific knowledge about the form of the A_n . In the same paper he shows that there exists a norming sequence $\{B_n\}$ with the stability property that $\lim_{n\to\infty} B_{n+1}B_n^{-1} = I$.

3. Differences in the infinite-dimensional case. The simplest situation in an infinite-dimensional Hilbert space occurs when there is an orthonormal basis along which all the projections of X have finite second moments.

Proposition 4. Suppose there exists an orthonormal basis $\{\tau_k\}$ such that $0 < \sigma_k^2 \equiv E\langle X, \tau_k\rangle^2 < \infty$ $\forall k$, then $X \in$ GDOA of a nondegenerate Gaussian.

Proof. Let $T_1 x = \sum_{k\geq 1} \lambda_k \sigma_k^{-1} \langle x, \tau_k\rangle e_k$ where $\{\lambda_k\}$ is a square summable sequence with $\sum(\lambda_k/\sigma_k)^2 < \infty$. Then $E\|T_1 X\|^2 < \infty$, hence there exists a nondegenerate $\mathcal{H}$-valued Gaussian W such that $\mathcal{L}(n^{-\frac{1}{2}} T_1 S_n) \to \mathcal{L}(W)$. In this case the norming sequence $T_n = n^{-\frac{1}{2}} T_1$ works. □

Random vectors satisfying the hypotheses of Proposition 4 may be rather hard to recognize. For instance, the projections of X along one orthonormal basis might all have infinite second moments while the projections along another might all have finite second moments. For a specific example consider

$$Y(x) = \sum (f_k(x)/\sqrt{g(x)})e_k$$

where $x \in \mathbb{R}$, g is a positive probability density and $f_k(x) = I_{[m_k, m_k']}(x) - I_{[-m_k', -m_k]}(x)$ for a sequence $0 \leq m_1 < m_1' < m_2 < m_2' < \ldots$ with $m_k' - m_k \geq 2^k$. Now $E\langle Y, e_k\rangle^2 = 2(m_k' - m_k) < \infty$ while if $a = \sum 2^{-k/2} e_k$ then

$$E\langle Y, a\rangle^2 = \sum 2^{-k} E\langle Y, e_k\rangle^2 \geq \sum 2^{-k} 2^{k+1} = \infty .$$

Hence, the Gram-Schmidt procedure applied to $\{a, e_1, e_2, \ldots\}$ yields an orthonormal basis $\{\tau_k\}$ such that $E\langle Y, \tau_k\rangle^2 = \infty$.

Perhaps the first natural question is whether only bounded linear operators occur as norming operators. After all, as we have seen, in $\mathbb{R}^k$ $\lim_{n\to\infty} \|T_n\| = 0$. The following example shows that unbounded linear operators may appear for norming and even when all the operators are bounded it is possible for $\lim_{n\to\infty} \|T_n\| = \infty$.

Example 1. Let X be a symmetric, nondegenerate $\mathcal{X}$-valued random vector with $E\|X\|^2 < \infty$. Thus, there exists a nondegenerate Gaussian W_1 with $\mathcal{L}(S_n/\sqrt{n}) \to \mathcal{L}(W_1)$. Define an unbounded linear operator A_1 by

$$A_1 x = \sum \sigma_k^{-1} b_k \langle x, e_k \rangle e_k$$

where $\sigma_k^2 = E\langle X, e_k\rangle^2$, b_k is a square summable sequence with $\sum \sigma_k^{-1} b_k = \infty$, and $\lim_{n\to\infty} b_n/\sigma_n\sqrt{n} = \infty$. The domain of A_1 , $\mathcal{D}(A_1)$, is the set of finite linear combinations of basis elements. Also, A_1 is closable. Now $E\|A_1 X\|^2 = \sum b_k^2 < \infty$, so there is a nondegenerate Gaussian W_2 such that if $A_n = A_1/\sqrt{n}$ then $\mathcal{L}(A_n S_n) \to \mathcal{L}(W_2)$. Consequently, the same sequence of partial sums S_n may be normalizable by both bounded and unbounded linear operators.

Restricting attention to bounded operators only, let $U_n = \Pi_n A_n + (I - \Pi_n)/\sqrt{n}$ where Π_n denotes projection on the subspace spanned by the first n basis elements. Since $\Pi_n x_n \to x$ whenever $x_n \to x$ in $\mathcal{X}$, $\mathcal{L}(\Pi_n A_n S_n) \to \mathcal{L}(W_2)$ while $\mathcal{L}((I - \Pi_n)S_n/\sqrt{n}) \to 0$. Therefore, $\mathcal{L}(U_n S_n) \to \mathcal{L}(W_2)$. However, $\lim_{n\to\infty} \|U_n\| \geq \lim_{n\to\infty} \|\Pi_n A_n\| \geq \lim_{n\to\infty} b_n/\sigma_n\sqrt{n} = \infty$. □

To further illustrate the differences between the finite and infinite-dimensional situations, the next example shows the failure in infinite dimensions of the following facts which we have seen to hold in $\mathbb{R}^k$ when $X \in$ GDOA :

(a) $\langle X, \theta\rangle \in$ DOA , $\forall\theta$;

(b) $\lim_{t\to\infty} \sup_{\|\theta\|=1} \frac{t^2 P(|\langle X, \theta\rangle| > t)}{E(|\langle X, \theta\rangle|^2 \wedge t^2)} = 0$;

(c) $E\|X\|^p < \infty$, $\forall p < 2$;

(d) $M_X(t)$ is a slowly varying function of t .

Example 2. Let $\mathcal{L}(X) = \sum \frac{c}{2k^2}[\delta_{a_k e_k} + \delta_{-a_k e_k}]$ where $\sum c/k^2 = 1$, $a_k = k2^{k/2}$, and δ_b is the measure with unit mass at b. Since $E\langle X, e_k\rangle^2 = c2^k$, Proposition 4 implies $X \in$ GDOA of a nondegenerate Gaussian.

If $\theta = \sum 2^{-k/2} e_k$, then

$$\mathcal{L}(\langle X,\theta\rangle) = \sum \frac{c}{2k^2}[\delta_k + \delta_{-k}] \quad \text{and}$$

$$j^2 P(|\langle X,\theta\rangle| > j)/E(|\langle X,\theta\rangle|^2 \wedge j^2)$$

$$= j^2 \sum_{k\geq j+1} ck^{-2}/(jc + j^2 \sum_{k\geq j+1} ck^{-2})$$

$$\geq j \int_{j+2}^{\infty} x^{-2}dx \Big/ \left(1 + j\int_j^{\infty} x^{-2}dx\right)$$

$$= j/2(j+2) \geq 1/4 \quad \text{if } j \geq 2 .$$

Consequently, $\langle X,\theta\rangle \notin$ DOA and (b) fails.

The failure of (c) is immediate because if $p \geq 1$,

$$E\|X\|^p \geq E|\langle X,\theta\rangle|^p = \sum ck^{p-2} = \infty .$$

Notice, however, that in this case there is a bounded linear operator T such that $E\|TX\|^p < \infty$ $\forall p \leq 2$; namely, $Tx = \sum (k2^{k/2})^{-1}\langle x, e_k\rangle e_k$.

The equivalence of (1) and (1') and the calculations in Proposition 2 show that $M_X(t)$ slowly varying implies $E\|X\|^p < \infty$, $\forall p < 2$. Consequently (d) fails for this example. □

3. Positive results in infinite dimensions. The previous example shows that the Hilbert space analogue of condition (3) is certainly not necessary for X to be in the GDOA of a nondegenerate Gaussian. The more general sufficient conditions of the next theorem are the Hilbert space analogues of conditions (6) and (7) which are crucial to the proof of Theorem 1.

Theorem 5. Let X be a symmetric, $\mathscr{H}$-valued, nondegenerate random vector. Suppose there exist a sequence of orthonormal bases $\{\tau_k^n\}$ and a sequence of positive decreasing numbers λ_k with $\sum \lambda_k \leq 1$ such that

(a) $\forall k \neq j$,

$$\lim_{n\to\infty} nE\langle X,\tau_k^n\rangle\langle X,\tau_j^n\rangle I_{(|\langle X,\tau_k^n\rangle|\leq a_n(\tau_k^n),\,|\langle X,\tau_j^n\rangle|\leq a_n(\tau_j^n))}/a_n(\tau_k^n)a_n(\tau_j^n)$$

$$= 0 ;$$

and

(b) $\forall \epsilon > 0$,

$$\lim_{n\to\infty} \sum_{k\geq 1} \frac{n\sqrt{\lambda_k}}{a_n(\tau_k^n)} E|\langle X,\tau_k^n\rangle| I_{(|\langle X,\tau_k^n\rangle| > \epsilon a_n(\tau_k^n))} = 0 .$$

Then, there exists a sequence of linear operators T_n such that $\mathscr{L}(T_nS_n) \to \mathscr{L}(W)$ where W is a nondegenerate Gaussian with a covariance operator S which has eigenvalues $\lambda_1^2, \lambda_2^2, \ldots$. Moreover, the T_n may be chosen to take the form

$$(10) \qquad T_n x = \sum (\lambda_k/a_n(\tau_k^n))\langle x,\tau_k^n\rangle e_k .$$

Proof. In order to simplify the notation we suppress the superscript n on the basis elements τ_j^n since the same proof works for both a constant and a varying basis. It suffices to verify the following three conditions which can be deduced from Theorem 4.3 of de Acosta, Araujo and Giné (1978) upon noticing that the set F of finite linear combinations of the standard basis elements is a w^*-dense subset of $\mathcal{X}$:

(i) $\forall \epsilon > 0 , \quad \lim_{n\to\infty} nP\{\|T_n X\| > \epsilon\} = 0 ;$

(ii) $\exists \delta > 0$ such that for all $y \in F$,

$$\lim_{n\to\infty} nE\langle T_n X I_{(\|T_n X\| \le \delta)}, y\rangle^2 = \langle Sy, y\rangle ;$$

(iii) $\lim_{N\to\infty} \sup_n \; n \sum_{k\ge N+1} E\langle T_n X I_{(\|T_n X\| \le 1)}, e_k\rangle^2 = 0 .$

Using, in order, $\sum \lambda_k \le 1$, $\sqrt{\lambda_k} \le 1$ and condition (b),

$$\begin{aligned} nP(\|T_n X\| > \epsilon) &= nP\Big(\sum_{k\ge 1} (\lambda_k/a_n(\tau_k))^2 \langle X, \tau_k\rangle^2 > \epsilon^2\Big) \\ &\le n \sum_{k\ge 1} P(\sqrt{\lambda_k}\, |\langle X, \tau_k\rangle| > \epsilon a_n(\tau_k)) \\ &\le \epsilon^{-1} \sum_{k\ge 1} \frac{n\sqrt{\lambda_k}}{a_n(\tau_k)} E|\langle X, \tau_k\rangle| I_{(|\langle X, \tau_k\rangle| > \epsilon a_n(\tau_k))} \\ &\to 0 \quad \text{as} \quad n \to \infty . \end{aligned}$$

Thus, (i) is verified.

The validity of (iii) follows from the defining property (4) of $a_n(\tau_k)$ and condition (b) as follows.

$$n \sum_{k\geq N+1} E\langle T_n X I_{(\|T_nX\|\leq 1)}, e_k\rangle^2$$

$$= n \sum_{k\geq N+1} E((\lambda_k/a_n(\tau_k))^2 \langle X, \tau_k\rangle^2 I_{(\|T_nX\|\leq 1)})$$

$$\leq \sum_{k\geq N+1} \lambda_k^2 + n \sum_{k\geq N+1} (\lambda_k/a_n(\tau_k))^2 E\langle X, \tau_k\rangle^2 I_{(\|T_nX\|\leq 1, |\langle X, \tau_k\rangle|>a_n(\tau_k))}$$

$$\leq \sum_{k\geq N+1} \lambda_k^2 + \sum_{k\geq N+1} \frac{n\lambda_k}{a_n(\tau_k)} E|\langle X, \tau_k\rangle| I_{(|\langle X, \tau_k\rangle|>a_n(\tau_k))}$$

$$\to 0 \quad \text{as} \quad N \to \infty .$$

Finally, in order to verify (ii), let $y = \sum_{k=1}^{m} b_k e_k$. Now

$$nE\langle T_nXI_{(\|T_nX\|\leq\delta)}, y\rangle^2 = nE\Big(\sum_{k=1}^{m} \frac{\lambda_k b_k}{a_n(\tau_k)} \langle X, \tau_k\rangle I_{(\|T_nX\|\leq\delta)}\Big)^2$$

$$= n\sum_{k=1}^{m} (\lambda_k b_k/a_n(\tau_k))^2 E\langle X, \tau_k\rangle^2 I_{(\|T_nX\|\leq\delta)}$$

$$+ n \sum_{k=1}^{m} \sum_{\substack{j=1\\ \neq k}}^{m} \frac{\lambda_k\lambda_j b_k b_j}{a_n(\tau_k)a_n(\tau_j)} E\langle X, \tau_k\rangle\langle X, \tau_j\rangle I_{(\|T_nX\|\leq\delta)} .$$

The behavior of these two terms, denoted Term I and Term II, will be considered separately.

Term I may be rewritten in a more convenient form as

$$\sum_{k=1}^{m} \lambda_k^2 b_k^2 nE\Big[\Big(\frac{\langle X, \tau_k\rangle^2}{a_n^2(\tau_k)} \wedge 1\Big)\Big(1 - I_{(\|T_nX\|>\delta)}\Big)\Big]$$

$$+ \sum_{k=1}^{m} \lambda_k^2 b_k^2 nE\Big(\frac{\langle X, \tau_k\rangle^2}{a_n^2(\tau_k)} - 1\Big) I_{(\|T_nX\|\leq\delta, |\langle X, \tau_k\rangle| > a_n(\tau_k))} .$$

The definition of $a_n(\tau_k)$ and the validity of (i) show that the first summand tends to $\sum_{k=1}^{m} \lambda_k^2 b_k^2$ as $n \to \infty$.

The second summand is dominated by

$$\|y\|^2 \sum_{k=1}^{m} \delta \frac{n\sqrt{\lambda_k}}{a_n(\tau_k)} E|\langle X, \tau_k\rangle| I_{(|\langle X, \tau_k\rangle| > a_n(\tau_k))} .$$

Hence, by an application of condition (b) the second summand vanishes in the limit. Thus, we may conclude that Term I converges to $\langle Sy, y\rangle$. This, of course, means that Term II must converge to zero.

Now |Term II|

$$= \left| \sum_{k=1}^{m} \sum_{\substack{j=1 \\ \neq k}}^{m} \frac{\lambda_k \lambda_j b_k b_j}{a_n(\tau_k) a_n(\tau_j)} n E\langle X, \tau_k\rangle\langle X, \tau_j\rangle I_{(\|T_n X\| \leq \delta)} \right|$$

$$\leq \sum_{k=1}^{m} \sum_{\substack{j=1 \\ \neq k}}^{m} \frac{\lambda_k \lambda_j |b_k b_j|}{a_n(\tau_k) a_n(\tau_j)} n \left| E\langle X, \tau_k\rangle\langle X, \tau_j\rangle I_{(\|T_n X\| \leq \delta)} \right|$$

$$= \sum_{k=1}^{m} \sum_{\substack{j=1 \\ \neq k}}^{m} \frac{\lambda_k \lambda_j |b_k b_j|}{a_n(\tau_k) a_n(\tau_j)} n \left| E\langle X, \tau_k\rangle\langle X, \tau_j\rangle [I_A - I_B + I_C + I_D + I_E] \right|$$

where $A = \{|\langle X, \tau_k\rangle| \leq a_n(\tau_k), |\langle X, \tau_j\rangle| \leq a_n(\tau_j)\}$

$B = A \cap \{\|T_n X\| > \delta\}$

$C = \{|\langle X, \tau_k\rangle| \leq a_n(\tau_k), |\langle X, \tau_j\rangle| > a_n(\tau_j), \|T_n X\| \leq \delta\}$

$D = \{|\langle X, \tau_k\rangle| > a_n(\tau_k), |\langle X, \tau_j\rangle| \leq a_n(\tau_j), \|T_n X\| \leq \delta\}$

$E = \{|\langle X, \tau_k\rangle| > a_n(\tau_k), |\langle X, \tau_j\rangle| > a_n(\tau_j), \|T_n X\| \leq \delta\}$.

Condition (a) implies that the term containing I_A converges to zero. The remaining terms are dominated by

$$\left(\sum_{k=1}^{m} \lambda_k |b_k|\right)^2 \quad nP(\|T_n X\| > \delta)$$

$$+ \|y\|^2 (2+\delta) \sum_{k=1}^{m} \frac{n}{a_n(\tau_k)} E|\langle X, \tau_k\rangle| I_{(|\langle X, \tau_k\rangle| > a_n(\tau_k))}$$

which converges to zero as a result of condition (b) and (i). Thus, (ii) is verified. □

It is easy to check that Example 2 satisfies (a) and (b) even though it does not satisfy (3). In the particular situation when $\sup_{\|\theta\|=1} 1/P(\langle X, \theta\rangle \neq 0) = M < \infty$, $a_n(\theta)$ are defined for all θ for $n \geq M$. So the proof in Hahn and Klass (1978) shows that $\theta \to a_n(\theta)$ is continuous. Consequently, just as in Hahn and Klass, (3) implies (6) and hence (b).

Furthermore, the reasoning mentioned in Section 2 shows that (a) is satisfied in the above case whenever there exists a sequence of preferred orthonormal bases $\{\tau_k^n\}$ constructed in exactly the same manner as in Theorem 1 from minimal norming constants. We have thus concluded the analogue of the finite-dimensional case, namely

<u>Corollary 6</u>. Let X be a symmetric, $\mathcal{H}$-valued nondegenerate random vector. Suppose $\sup_{\|\theta\|=1} 1/P(\langle X, \theta\rangle \neq 0) < \infty$. Assume (3) and the existence of a sequence of orthonormal bases constructed using minimal norming constants. Then $X \in$ GDOA of a nondegenerate Gaussian.

It is clear that an orthonormal basis cannot always be constructed using minimal norming constants, even when X has independent coordinates. For example, let $Y, Y_1, Y_2, \ldots$ be i.i.d. symmetric random variables with $EY^2 < \infty$. Define $X = \sum k^{-1} Y_k e_k$. Now $E\langle X, e_k\rangle^2 < \infty$ for all k, so $X \in$ GDOA of a nondegenerate Gaussian. However, $a_n(e_k) = k^{-1} a_n(e_1)$, so $\inf_{\|\theta\|=1} a_n(\theta) = 0$.

Concerning the sharpness of Theorem 1, we know that condition (a) is necessary whenever norming is possible by a sequence taking the form in (10). We suspect that condition (b) is too strong.

Proposition 7. Suppose $X \in$ GDOA of a nondegenerate Gaussian W with covariance operator $Sx = \sum \lambda_k^2 \langle x, e_k\rangle e_k$. If the norming linear operators can be chosen to take the form (10) then (a) holds.

Proof. Again, for notational convenience, we suppress the superscript n. According to Theorem 2.12 of de Acosta, Araujo and Giné (1978), two necessary conditions are, $\forall \delta > 0$,

(i) $\lim_{n\to\infty} nP(\|T_n X\| > \delta) = 0$,

(ii) $\lim_{\delta \downarrow 0} \lim_{n\to\infty} nE\langle T_n X I_{(\|T_n X\| \leq \delta)}, y\rangle^2 = \langle Sy, y\rangle$.

Because of (i), (ii) can be replaced by

(ii') $\lim_{n\to\infty} nE\langle T_n X I_{(\|T_n X\| \leq 1)}, y\rangle^2 = \langle Sy, y\rangle$.

Letting $y = e_i + e_j$,

$$nE\langle T_nXI_{(\|T_nX\|\leq 1)}, e_i+e_j\rangle^2 = n\sum_{k=i,j} \lambda_k^2 E\langle X,\tau_k\rangle^2 I_{(\|T_nX\|\leq 1)}/a_n(\tau_k)$$

$$+ 2n\lambda_i\lambda_j E\langle X,\tau_i\rangle\langle X,\tau_j\rangle[I_A - I_B + I_C + I_D + I_E]/a_n(\tau_i)a_n(\tau_j)$$

where A through E are as defined in the proof of Theorem 5. If we can show that the first term converges to $\langle Sy, y\rangle$, then since the second summand with I_A deleted is $O(nP(\|T_nX\| > 1))$ which goes to zero by (i), condition (ii') yields the necessity of condition (a).

Rewriting the first term and using the definition of $a_n(\tau_k)$,

$$n \sum_{k=i,j} \lambda_k^2 [E(\frac{\langle X,\tau_k\rangle^2}{a_n^2(\tau_k)} \wedge 1) - P(|\langle X,\tau_k\rangle| > a_n(\tau_k)) +$$

$$E\langle X,\tau_k\rangle^2 (I_{(|\langle X,\tau_k\rangle| > a_n(\tau_k), \|T_nX\|\leq 1)} - I_{(|\langle X,\tau_k\rangle|\leq a_n(\tau_k), \|T_nX\| > 1)})/a_n^2(\tau_k)]$$

$$= \sum_{k=i,j} \lambda_k^2 + O(nP(\|T_nX\| > 1)) \to \sum_{k=i,j} \lambda_k^2 = \langle Sy, y\rangle \text{ by (i) .}$$

Therefore condition (a) is necessary. □

REFERENCES

de Acosta, A., Araujo, A. and Giné, E. (1978). On Poisson measures, Gaussian measures and the central limit theorem in Banach space. (Preprint)

Billingsley, P. (1966). Convergence of types in k-space. Z. Wahrscheinlichkeitstheorie 5, 175-179.

Hahn, M. G. and Klass, M. J. (1978). Matrix normalization of sums of i.i.d. random vectors in the domain of attraction of the multivariate normal (to appear in Ann. Probability).

Urbanik, K. (1972). Lévy's probability measures on Euclidean spaces. Studia Math. 44, 119-148.

FOOTNOTE

1. This research was partially supported by National Science Foundation Grant MCS-78-02417.

RELATION BETWEEN CENTRAL-LIMIT THEOREM AND LAW OF THE ITERATED LOGARITHM IN BANACH SPACES

B. HEINKEL

Since the beginning of the study of the limit properties of Banach space valued r.v., the following question raised :
"X being a Banach space valued r.v., satisfying the central-limit theorem, under what conditions does it satisfy also the law of the iterated logarithm?'

We are going to study this problem, in particular for some r.v. having a non square-integrable norm, in which situation the classical results (see Theorems 1 and 2 below) don't allow to conclude. Before to give the result, we introduce some notations .

We only consider centered r.v. which take their values in a real separable Banach space $(B, \|.\|)$ equipped with its Borel σ-field $\mathcal{B}$.

Let X be such a r.v. , and let $(X_n)_{n \in \mathbb{N}}$ denote a sequence of independent copies of X . For all $n \in \mathbb{N}$, we put :

$$S_n(X) = \sum_{k=1}^{n} X_k$$

Let L_2 denote the following function :

$$L_2 : \mathbb{R}^+ \to \mathbb{R}^+$$

$$L_2(x) = \mathrm{Log}(\mathrm{Log}\ x) \qquad \forall\, x \geq e^e$$

$$= 1 \qquad \forall\, x \in [o, e^e[$$

For simplication, we put :

$$\forall\, n \in \mathbb{N} \qquad a_n = \sqrt{2nL_2\, n}$$

Let us recall the following definition :

DEFINITION .- Let us consider $X : (\Omega,\mathcal{F},P) \to (B,\mathcal{B})$, $EX = 0$.

1) X satisfies the central-limit theorem (what we denote by "CLT") if $\left(\frac{S_n(X)}{\sqrt{n}}\right)_{n\in\mathbb{N}}$ converges in distribution as $n \to +\infty$.

2) X satisfies the law of the iterated logarithm (what we denote by "LIL") if $P\{\left(\frac{S_n(X)}{a_n}\right)_{n\in\mathbb{N}}$ is $\|.\|$ - conditionally compact$\} = 1$.

The classical results which precise the relations between these two properties are the following :

THEOREM 1 : (G. Pisier [4] Théorème 4.3)

Consider $X : (\Omega,\mathcal{F},P) \to (B,\mathcal{B})$, $EX = 0$ with $E\|X\|^2 < +\infty$, satisfying the CLT . Then X also satisfies the LIL .

THEOREM 2 : (J. Kuelbs [2] Theorem 4.1)

Consider $X : (\Omega,\mathcal{F},P) \to (B,\mathcal{B})$, $EX = 0$ with $E\|X\|^2 < +\infty$, such that :

$$\frac{S_n(X)}{a_n} \xrightarrow[n\to+\infty]{P} 0$$

Then X also satisfies the LIL .

A well known necessary condition for X to satisfy the LIL being :

$$E\,\frac{\|X\|^2}{L_2\|X\|} < +\infty$$

the following problem remains : "For what r.v. X with $E\|X\|^2 = +\infty$, $E\frac{\|X\|^2}{L_2\|X\|} < +\infty$, does the CLT imply the LIL ? ".

A partial answer to this question is brought by :

THEOREM 3 : Consider $X : (\Omega,\mathcal{F},P) \to (B,\mathcal{B})$, $EX = 0$, such that :

1) X satisfies the CLT .

2) $\exists\ \varepsilon \in \,]0,1[$: $E\,\frac{\|X\|^2}{(L_2\|X\|)^\varepsilon} < +\infty$

Then X also satisfies the LIL .

By Pisier's theorem giving the LIL in terms of approximations by simple functions (cf [4] Théorème 3.1) , it is easy to check that it is sufficient to prove the following :

PROPOSITION : Consider $X : (\Omega, \mathcal{F}, P) \to (B, \mathcal{B})$, $EX = 0$, such that :

1) $\sup_n E \frac{\|S_n(X)\|}{\sqrt{n}} < +\infty$

2) $\exists\, \varepsilon \in]0,1[\; : \; E \frac{\|X\|^2}{(L_2\|X\|)^\varepsilon} < +\infty$

Then, we also have : $E \sup_n \frac{\|S_n(X)\|}{a_n} < +\infty$

It is well known that the proof of the Proposition reduces to show :

$$\sup_n \frac{\|S_n(X)\|}{a_n} < +\infty \qquad \text{a.s.}$$

Furthermore we can suppose that X is symmetric (the general case easily follows) and that $\varepsilon = \frac{\alpha}{\alpha+1}$, $\alpha \in \mathbb{N}$, what is not a loss of generality .

Now, we are proceeding by truncation. We put :

$$a_0^n = n^{\frac{1}{2}} (L_2 n)^{-\frac{1}{2}}$$

$$a_1^n = n^{\frac{1}{2}} (L_2 n)^{-\frac{\varepsilon}{2}}$$

$$a_2^n = n^{\frac{1}{2}} (L_2 n)^{\frac{1-2\varepsilon}{2}}$$

$$\vdots$$

$$a_k^n = n^{\frac{1}{2}} (L_2 n)^{\frac{(k-1)-k\varepsilon}{2}}$$

$$\vdots$$

$$a_{2\alpha+1}^n = n^{\frac{1}{2}} (L_2 n)^{\frac{\varepsilon}{2}}$$

and :

$$\eta_o^n = X_n \, I_{\{\|X_n\| < a_o^n\}}$$

$$\eta_1^n = X_n \, I_{\{a_o^n \le \|X_n\| < a_1^n\}}$$

$$\vdots$$

$$\eta_k^n = X_n \, I_{\{a_{k-1}^n \le \|X_n\| < a_k^n\}}$$

$$\vdots$$

$$\eta_{2\alpha+2}^n = X_n \, I_{\{\|X_n\| \ge a_{2\alpha+1}^n\}}$$

The choices of a_o^n and $a_{2\alpha+1}^n$ are explained by the following two properties :

i) $P\ \{\sup_1^\infty \|\sum_{j=1}^n \frac{\eta_o^j}{a_n}\| < +\infty\} = 1$

([4] Proposition 4.3)

ii) $\|\sum_{j=1}^n \frac{\eta_{2\alpha+2}^j}{a_n}\| \xrightarrow[n\to+\infty]{a.s.} 0$

which follows easily from the hypothesis 2) by the Borel-Cantelli Lemma .

It remains to check :

iii) $\forall\ k = 1,2\ ,\ldots,\ 2\alpha+1$

$$\|\sum_{j=1}^n \frac{\eta_k^j}{a_n}\| \xrightarrow[n\to+\infty]{a.s.} 0$$

The proof of property iii) is based on the Kuelbs-Zinn truncation method [3]. I'm only giving a sketch of the proof. Suppose k fixed $(=1,2,\ldots,2\alpha+1)$ and put for simplication :

$$\forall j \in \mathbb{N} \quad \eta_j = \eta_k^j$$

$$I(j) = \{2^j + 1,\ldots,\ 2^{j+1}\}$$

It is easy to see that, in fact, it suffices to show :

$$\left\| \sum_{j\in I(n)} \frac{\eta_j}{a_{2^n}} \right\| \xrightarrow[n\to+\infty]{a.s.} 0$$

If we note :

$$\Lambda(n) = \sum_{j\in I(n)} \frac{E\|\eta_j\|^2}{(a_{2^{n+1}})^2}$$

we remark that :

$$\sum_{n=1}^{\infty} \Lambda(n) \leq c \ E\frac{\|X\|^2}{(L_2\|X\|)^{\varepsilon}}$$

and this is the stage of the proof where we need the particular form of the a_k^n

Let $\delta > 0$ be given ; for all $j \in I(n)$ we put :

$$h_j = \eta_j \, I_{\{\|\eta_j\| \leq \Lambda(n)^{\frac{1}{4}} a_{2^{n+1}}\}}$$

$$k_j = \eta_j \, I_{\{\|\eta_j\| > \frac{\delta}{2} a_{2^{n+1}}\}}$$

$$l_j = \eta_j - k_j - h_j$$

Furthermore :

$$U_n^1 = \left\| \sum_{j\in I(n)} h_j \right\|$$

$$U_n^2 = \left\| \sum_{j\in I(n)} k_j \right\|$$

$$U_n^3 = \left\| \sum_{j\in I(n)} l_j \right\|$$

The proof of iii) and at the same time the one of the Proposition are finished by showing :

$\forall \delta > 0 \ \ \forall j = 1,2,3$ the following property is realized :

$$(j) \quad \sum_{n=1}^{\infty} P\{U_n^j > \delta a_{2^{n+1}}\} < +\infty$$

This is done in the same way as in the proof of Theorem 1 in [3] .

Some concluding remarks .

It can be shown by refining the calculations ([1]) that the condition (2) of the Theorem 3 can be weakened in :

$$E \frac{\|X\|^2 L_3 \|X\|}{L_2 \|X\|} < +\infty$$

(where L_3 is defined in the same way as L_2) .

This leads to the following naturel questions :

1) Describe the class F of increasing functions $f : \mathbb{R}^+ \to \mathbb{R}^+$, such that :

a) $\lim\limits_{x \to +\infty} \frac{f(x) L_2(x)}{x^2 L_3(x)} = 0$

b) If X satisfies the CLT and $Ef(\|X\|) < +\infty$, then X also satisfies the LIL .

2) Does F depend on the space B where X takes its values ?

3) If the answer to question 2 is positive, for what Banach spaces B does $x \to \frac{x^2}{L_2(x)}$ belong to F ?

REFERENCES

[1] B. HEINKEL : Sur la relation entre théorème central-limite et loi du logarithme itéré dans les espaces de Banach . (to appear)

[2] J. KUELBS : Kolmogorov law of the iterated logarithm for Banach space valued random variables . Illinois Journal of Maths. 21-4 (1977), 784-800 .

[3] J. KUELBS, J. ZINN : Some stability results for vector valued random variables . Preprint .

[4] G. PISIER : Le théorème de la limite centrale et la loi du logarithme itéré dans les espaces de Banach . Séminaire Maurey-Schwartz 1975-76 , exposés n° 3 et 4 .

INSTITUT DE RECHERCHE MATHEMATIQUE AVANCEE
Laboratoire Associé au C.N.R.S.
Université Louis Pasteur
7, rue René Descartes
67084 - STRASBOURG Cedex

RATES OF GROWTH FOR BANACH SPACE VALUED INDEPENDENT INCREMENT PROCESSES

J. Kuelbs*
University of Wisconsin
Department of Mathematics
Madison, WI 53706/USA

1. Introduction.

Let $X = \{X(t): t \geq 0\}$ be a separable stochastically continuous process with stationary independent increments which takes values in a real separable Banach space B having norm $\|\cdot\|$. In the special case where X is a B-valued Brownian motion and $|\cdot|$ is a continuous seminorm on B the rate of growth of

$$(1.1) \qquad \sup_{s \leq t} |X(s)|$$

as $t \to \infty$ was studied in [19], and that of

$$(1.2) \qquad \inf_{s \geq t} |X(s)|$$

as $t \to \infty$ (thus giving the rate of escape of X) in [12].

In this article we will examine (1.1) and (1.2) for X as described above and present some examples demonstrating the substantial difference between the infinite and finite dimensional cases. In fact, in [19] integral tests for upper and lower functions were obtained for Brownian motion in B, and some examples when $B = \mathbf{R}^d$, or even $\mathbf{R}^2$, demonstrate that the classical results on the rate of growth of (1.1) depend greatly on the norm being nearly Euclidean. Now the Dvoretsky-Erdös test [11] for the rate of escape of Brownian motion in $\mathbf{R}^d$ is invariant under equivalent norms, and hence determining the rate of escape of a Brownian motion with infinite dimensional state space B naturally presented itself as an intriguing problem at the time [19] was written. However, nothing was accomplished along these lines until the problem was mentioned to Professor K. B. Erickson in the summer of 1977. His results and some joint examples are soon to appear in [12].

In Theorem 2 we present a strengthened form of Erickson's result on Brownian motion which provides information regarding (1.2) for stable processes. In Theorem 3 we obtain information regarding (1.1) for general independent increment processes and Theorem 4 applies this to stable processes. More important, however, are the examples involving (1.2) which are given in Section 3 and Section 4.

The examples of Section 3 deal with Brownian motion as well as with stable processes of any index. A comparison of Theorems 5 and 6 indicates that in the infinite-dimensional setting the rate of escape for symmetric stable processes depends less on α then is the case in finite dimensions. Furthermore, these

*Supported in part by NSF Grant MCS-77-01098.

examples demonstrate that some infinite dimensional processes have a "natural rate of escape".

In Section 4 an application of Theorem 2 is made to the Brownian sheet and the tied down Brownian sheet. Combining these results with some approximation results of Kiefer, and Berkes and Philipp, we obtain the rate of escape for the empirical process as well.

2. The rate of growth of $\inf_{s \geq t} |X(s)|$ and $\sup_{s \leq t} |X(s)|$.

Let $X = \{X(t) : t \geq 0\}$ be a separable B-valued process with stationary independent increments such that $X(0) = 0$. If the time parameter t is continuous we assume the process is stochastically continuous, and hence by [15, pp. 183-184] we can assume the sample paths are right continuous and have left hand limits with probability one. An additional important property of such a process is this: If T is any stopping time relative to $\{\sigma(X(s) : s \leq t) : t \geq 0\}$, then, conditional on $T < \infty$, the process $\{X(T+t) - X(T) : t \geq 0\}$ is independent of $\sigma(X(s) : s \leq T)$ and is a probabilistic replica of X. As in the introduction we are assuming B is a real separable Banach space with norm $\|\cdot\|$ and $|\cdot|$ is always a continuous semi-norm on B. Further, we assume $\gamma(t)$ is a positive non-decreasing function on $[0, \infty)$ which varies regularly at ∞ with positive exponent, i.e. for some $\beta > 0$ (the exponent) we have

$$\lim_{t \to \infty} \frac{\gamma(tx)}{\gamma(t)} = x^{\beta} \quad (x > 0). \tag{2.1}$$

The rate of escape for $X = \{X(t) : t \geq 0\}$ can be obtained from the following theorem due to Erickson [12].

<u>Theorem 1.</u> Let X and $\gamma(t)$ be as described above and assume $b > 1$. Then (2.2) implies (2.3) implies (2.4) where

$$\underline{\lim}_{t \to \infty} |X(t)|/\gamma(t) < 1 \quad \text{w.p. } 1, \tag{2.2}$$

$$\sum_k P(|X(t)| \leq \gamma(t) \text{ for some } t \in [b^k, b^{k+1})) = \infty \tag{2.3}$$

$$\underline{\lim}_{t \to \infty} |X(t)|/\gamma(t) \leq 1 \quad \text{w.p. } 1, \tag{2.4}$$

In particular, when X is a sample continuous B-valued Brownian motion Theorem 1 can be applied to obtain an extension of the Dvoretsky-Erdös result [11] as was done in [12] by Erickson. To describe this result, as well as its extension to stable processes, we first need some terminology.

We say a probability measure μ is non-degenerate if μ is not concentrated at a single point. A non-degenerate probability measure μ on B is called stable if for every $a > 0$, $b > 0$, there exists a $\gamma > 0$ and $s \in B$ such that: if X and Y are independent random variables with $\mathcal{L}(X) = \mathcal{L}(Y) = \mu$, then

$\mathcal{L}(aX + bY) = \mathcal{L}(\gamma X + s)$. The measure μ is said to be strictly stable if for every $a > 0$, $b > 0$ the choice of $s = 0$ is possible. For every stable measure on the Borel sets of B it can be proved that there exists a number $\alpha \in (0, 2]$ such that if a, b, γ are as above, then $\gamma = (a^\alpha + b^\alpha)^{1/\alpha}$ is possible [10, Theorems 4 and 5]. The number $\alpha \in (0,2]$ is called the index of μ, and is unique since μ is assumed to be non-degenerate. If μ has index 2, then every linear functional has a Gaussian distribution and hence μ is a Gaussian measure on B.

If $X = \{X(t) : t \geq 0\}$ is a stochastically continuous separable B-valued process with stationary independent increments, $X(0) = 0$, and $\mathcal{L}(X(1)) = \mu$ where μ is a strictly stable probability on B of index α, then we call X a strictly stable process of index α. In view of the scaling property of strictly stable measures the process X satisfies the scaling property $\mathcal{L}(X(t)) = \mathcal{L}(t^{1/\alpha}X(1))$.

Using the terminology of [12] we say a semi-norm $|\cdot|$ on B is of rank at least d with respect to $\{X(t) : t \geq 0\}$ if there is a continuous linear transformation $\Lambda : B \to B$ of rank d such that the process $\{\Lambda X(t) : t \geq 0\}$ is genuinely d-dimensional and

$$\|\Lambda x\| \leq |x| \quad (x \in B).$$

Finally, we call a function h defined on $[0, \infty)$ admissible of order α if (i) $0 < h(t) \searrow 0$ and $t^{1/\alpha}h(t) \nearrow \infty$ as $t \nearrow \infty$, and (ii) h varies slowly at infinity.

The next theorem includes Erickson's extension of the Dvoretsky-Erdös result [11] for Brownian motion in $\mathbf{R}^d$.

<u>Theorem 2.</u> Let $X = \{X(t) : t \geq 0\}$ be a stochastically continuous separable B-valued process with stationary independent increments, $X(0) = 0$, and assume X is strictly stable of index α. If $|\cdot|$ is a continuous semi-norm with dimension greater than α with respect to X, and $h(t)$ is admissible of order α, then

$$\varliminf_{t \to \infty} \frac{|X(t)|}{t^{1/\alpha} h(t)} \leq 1 \quad (\geq 1) \tag{2.5}$$

according as

$$\sum_{k=1}^{\infty} P(|X(1)| \leq h(b^k))h^{-\alpha}(b^k) = \infty \quad (< \infty). \tag{2.6}$$

Proof. In case $\alpha = 2$, this is a result due to Erickson [12].

For $\alpha < 2$, let $\gamma(t) = t^{1/\alpha} h(t)$. Then Theorem 1 applies with $\beta = \frac{1}{\alpha}$ in (2.1) by applying the following lemma.

<u>Lemma 1.</u> For any $b > 1$, $\varepsilon > 0$, and $k = 1, 2, \ldots$

$$a_1 h^{-\alpha}(b^k) P(|X(1)| \le h(b^k)) \le P(|X(s)| \le \gamma(s) \text{ for some } s \in [b^{k-1}, b^k))$$
$$\le a_2 h^{-\alpha}(b^{k-1}) P(|X(1)| \le (1+\varepsilon) h(b^{k-1})).$$

where

$$a_1 = \left(\int_0^\infty P(|X(s)| \le 2)\, ds\right)^{-1} (1 - b^{-1}) > 0$$

and

$$a_2 = \max\left\{\frac{2h^\alpha(b)}{(b^2 - b)} \quad (b^2 - 1),\ 8(b^2 - 1)\varepsilon^{-\alpha}[E|X(1)|^{\alpha/2}]^2\right\} < \infty .$$

Proof. Exactly as the corresponding result in Erickson [12]. To see $a_1 > 0$ note that if $|\cdot|$ has dimension d, then by the scaling property

$$P(|X(s)| \le 2) = P(|X(1)| \le 2/s^{1/\alpha})$$
$$\le P(\|\Lambda X(1)\| \le 2/s^{1/\alpha})$$
$$\le c(2/s^{1/\alpha})^d$$

for some $c < \infty$. If $d > \alpha$, then $a_1 > 0$ as required.

To verify $a_2 < \infty$, we need $E|X(1)|^{\alpha/2} < \infty$ and this follows from the tail behavior of a stable law of index α [8].

remark 1. Let $B = \mathbb{R}^d$ with $|\cdot|$ denoting the usual Euclidean norm. Assume X is a symmetric stable process of index α which is genuinely d-dimensional, and let $h(t)$ be an admissible function of order α. Then, since $\mathcal{L}(X(1))$ has a density which is positive in a neighborhood of zero we have a positive constant $c(\alpha, d)$ such that

$$P(|X(1)| \le h(b^k)) \sim c(\alpha, d) h^d(b^k) \tag{2.7}$$

as $k \to \infty$. Hence (2.6) converges or diverges according as

$$\sum_{k=1}^{\infty} h^{d-\alpha}(b^k) \left(\int_1^\infty \frac{h^{d-\alpha}(s)\, ds}{s}\right) \tag{2.8}$$

converge's or diverges. Therefore, for the class of increasing functions of the form $\gamma(t) = t^{1/\alpha} h(t)$, Theorem 2 contains the Dvoretsky-Erdös result [11] provided $\alpha = 2$ and the result of J. Takeuchi [21, Theorem 1] when $\alpha < 2$.

remark 2. In the finite dimensional case as described in the previous remark (2.5) is easily seen to be either 0 or ∞ by using (2.8). This is a result of (2.7) where we see that $P(|X(1)| \le \varepsilon)$ decays as a polynomial as $\varepsilon \downarrow 0$. In some infinite dimensional examples given in the next section we will obtain a nontrivial lim inf in (2.5). This, of course, results from the exponential character of $P(|X(1)| \le \varepsilon)$ as $\varepsilon \downarrow 0$ in the infinite dimensional setting.

Now we turn to the rate of growth of $\sup_{s \le t} |X(s)|$. The next theorem is similar to a number of known results in the finite dimensional setting, but its exact analogue is unknown to me. Its proof involves only fairly standard methods.

Theorem 3. Let $X = \{X(t) : t \ge 0\}$ and $\gamma(t)$ be as described prior to Theorem 1. Further, assume for every $\varepsilon > 0$ there exists $\delta > 0$ such that

$$\overline{\lim_{t \to \infty}} \; P(|X(t)| \ge \varepsilon\gamma(t)) \le 1 - \delta . \tag{2.9}$$

Then, with probability one

$$\overline{\lim_{t \to \infty}} \; \frac{|X(t)|}{\gamma(t)} \le 1 \qquad (\ge 1) \tag{2.10}$$

according as

$$I(\gamma) = \int_1^\infty \frac{P(|X(t)| > \gamma(t))}{t} \, dt < \infty \quad (= \infty). \tag{2.11}$$

Proof. First note that $\lim_{t \to \infty} \gamma(t) = \infty$ since $\beta > 0$. Now fix $\varepsilon > 0$, $\varepsilon < \frac{1}{4}$, and assume $I(\gamma) < \infty$. Take $b > 1$, $b < 2$, and define

$$E_k = \{ \sup_{t \le b^k} |X(t)| > (1 + 2\varepsilon)\gamma(b^{k+1}) \} .$$

Then (2.9), along with the stationarity of the increments and the separability of X, yields in a standard manner (see [5, p. 45] for details) that for sufficiently large k we have

$$P(E_k) \le \frac{1}{\delta} P(|X(b^k)| > (1 + \varepsilon)\gamma(b^{k+1})) . \tag{2.12}$$

Now for $t \in [b^k, b^{k+1}]$ we have $t - b^k \le (b-1)b^k \le b^k$, and hence by the stationarity of the increments, (2.2), and $\lim_{t \to \infty} \gamma(t) = \infty$ we have

$$P(|X(t) - X(b^k)| \ge \varepsilon\gamma(b^k)) \le 1 - \delta$$

for all sufficiently large k. Hence for $t \in [b^k, b^{k+1}]$ we have

$$\begin{aligned} P(E_k) &\le \frac{1}{\delta^2} P(|X(b^k)| > (1 + \varepsilon)\gamma(b^{k+1}))P(|X(t) - X(b^k)| \le \varepsilon\gamma(b^{k+1})) \\ &\le \frac{1}{\delta^2} P(|X(t)| > \gamma(t)) \end{aligned}$$

since $\gamma(t)$ in non-decreasing. Thus

$$P(E_k) \le \frac{1}{\delta^2} \frac{b}{b-1} \int_{b^k}^{b^{k+1}} P(|X(t)| > \gamma(t))/t \, dt ,$$

and since $I(\gamma) < \infty$ we have $\sum_k P(E_k) < \infty$.

Recall $\varepsilon > 0$, $\varepsilon < \frac{1}{4}$. Now $\gamma(t)$ non-decreasing and slowly varying with positive exponent $\beta > 0$ implies there exists $g > 2$ sufficiently large so that

$$\gamma(s_2)/\gamma(s_1) < 1 + \varepsilon/2 \quad \text{when} \quad g \le s_1 \le s_2 \le s_1 g/g-1 , \tag{2.13}$$

$$\gamma(s_1)/\gamma(s_2) < \varepsilon/2 \quad \text{when} \quad g \le s_1 \le s_2/g-1 . \tag{2.14}$$

To prove (2.13) and (2.14) use [4] since $\gamma(t)/t^{\beta}$ is slowly varying.

Hence for $b^2 = g/g-1$ we have for all k sufficiently large that $b^{k-1} \geq g$ and hence (2.13) and $0 < \varepsilon < \frac{1}{4}$ implies

$$(1+3\varepsilon)\gamma(b^{k-1}) \geq (1+2\varepsilon)\gamma(b^{k+1}).$$

Thus

$$\sum_k P\left(\sup_{t \leq b^k} |X(t)| > (1+3\varepsilon)\gamma(b^{k-1})\right) < \infty,$$

and since $\varepsilon > 0$ was arbitrary we have with probability one that

$$\overline{\lim_{t \to \infty}} \; |X(t)|/\gamma(t) \leq 1.$$

Now fix $\varepsilon > 0$, $\varepsilon < \frac{1}{4}$, and assume $I(\gamma) = \infty$. Using (2.13) with $b = g/_{g-1} > 1$ we have for all k sufficiently large and $s \in [b^k, b^{k+1}]$ that

$$\gamma(s) \geq (1-\varepsilon)\gamma(b^{k+1}). \tag{2.15}$$

Following the argument indicated for (2.12) we have for all sufficiently large k that

$$P\Big(\sup_{s \leq b^{k+1}} |X(s)| > (1-\varepsilon)\gamma(b^{k+1})\Big) \leq \frac{1}{\delta} P(|X(b^{k+1})| > (1-2\varepsilon)\gamma(b^{k+1})), \tag{2.16}$$

and hence for $b^k \leq s \leq b^{k+1}$ (2.15) implies

$$\begin{aligned} P(|X(b^{k+1})| > (1-2\varepsilon)\gamma(b^{k+1})) &\geq \delta P(|X(s)| > (1-\varepsilon)\gamma(b^{k+1})) \\ &\geq \delta P(|X(s)| > \gamma(s)), \\ &\geq \frac{\delta}{b-1} \int_{b^k}^{b^{k+1}} \frac{P(|X(s)| > \gamma(s))}{s}\, ds. \end{aligned}$$

Since $I(\gamma) = \infty$ we thus have for $b = g/_{g-1} > 1$ that

$$\sum_k P(|X(b^{k+1})| > (1-2\varepsilon)\gamma(b^{k+1})) = \infty. \tag{2.17}$$

Now (2.17) implies for every N there exists an ℓ such that

$$\sum_k P(|X(b^{\ell+Nk})| > (1-2\varepsilon)\gamma(b^{\ell+Nk})) = \infty. \tag{2.18}$$

Next we define the independent events

$$F_k = \left\{\frac{|X(b^{\ell+N(k+1)}) - X(b^{\ell+Nk})|}{1-2\varepsilon} > \gamma(b^{\ell+N(k+1)}) - \gamma(b^{\ell+Nk})\right\}$$

where N is to be specified later. Now by (2.9) there exists $\delta' > 0$ such that

$$P\left(\frac{|X(t)|}{1-2\varepsilon} \leq \gamma(t)/_2\right) \geq \delta'$$

for all sufficiently large t. Hence for all sufficiently large k

$$P(F_k) \geq \int_{\frac{|u|}{1-2\varepsilon} \leq \frac{\gamma(b^{\ell+Nk})}{2}} P\left(\frac{|X(b^{\ell+N(k+1)})-u|}{1-2\varepsilon} \geq \gamma(b^{\ell+N(k+1)}) - \gamma(b^{\ell+Nk})\right)$$

$$(2.19) \qquad \cdot P(X(b^{\ell+Nk}) \in du)$$

$$\geq \delta' P(|X(b^{\ell+N(k+1)})| \geq \gamma(b^{\ell+N(k+1)}) \; (1-2\varepsilon)) .$$

Hence (2.18) and (2.19) imply $\sum_k P(F_k) = \infty$ and by the Borel-Cantelli lemma $P(F_k \text{ i.o.}) = 1$.

Now choose N such that $b^N > g > 2$. Then (2.14) implies

$$(2.20) \qquad \gamma(b^{\ell+N(k+1)}) - 2\gamma(b^{\ell+Nk}) > (1-\varepsilon)\gamma(b^{\ell+N(k+1)}) .$$

Hence the event

$$D_k = \left\{\frac{|X(b^{\ell+N(k+1)})|}{1-2\varepsilon} > \gamma(b^{\ell+N(k+1)}) - 2\gamma(b^{\ell+Nk})\right\}$$

implies the event

$$G_k = \left\{\frac{|X(b^{\ell+N(k+1)})|}{1-2\varepsilon} > (1-\varepsilon)\gamma(b^{\ell+N(k+1)})\right\} .$$

Further, the event F_k implies either

$$C_k = \left\{\frac{|X(b^{\ell+Nk})|}{1-2\varepsilon} > \gamma(b^{\ell+Nk})\right\}$$

or D_k (and hence G_k). Thus $P(F_k \text{ i.o.}) = 1$, and $\{C_k \text{ i.o.}\}$ and $\{G_k \text{ i.o.}\}$ tail events, implies $P(C_k \text{ i.o.}) = 1$ or $P(G_k \text{ i.o.}) = 1$. In either case we have with probability one that

$$\overline{\lim_k} \; |X(b^{\ell+Nk})| / \gamma(b^{\ell+Nk}) \geq (1-2\varepsilon)(1-\varepsilon),$$

and since $\varepsilon > 0$ was arbitrary the theorem is proved.

As an easy application of Theorem 3 we can prove a result which is well known if $B = \mathbf{R}^d$.

<u>Theorem 4.</u> Let $X = \{X(t) : t \geq 0\}$ be a stochastically continuous separable B-valued process with stationary independent increments, $X(0) = 0$, and assume X is strictly stable of index $\alpha < 2$. If $h(t)$ is slowly varying at infinity and $\lim_{t\to\infty} h(t) = \infty$, then

$$(2.21) \qquad \overline{\lim_{t\to\infty}} \frac{|X(t)|}{t^{1/\alpha} h(t)} = 0 \qquad (+\infty)$$

according as

$$(2.22) \qquad \int_1^\infty \frac{1}{t[h(t)]^\alpha}\, dt < \infty \qquad (= \infty) .$$

Proof. First observe that by [8] we have a positive constant c such that $P(|X(1)| > t) \sim c/t^\alpha$ as $t \to \infty$. Hence by strict stability we have

$$P(|X(t)| > t^{1/\alpha} h(t)) = P(|X(1)| > h(t)) \sim \frac{c}{[h(t)]^\alpha} . \tag{2.23}$$

Thus $\lim_{t \to \infty} h(t) = \infty$ implies (2.9) with $\delta = 1$ and if $\gamma(t) = t^{1/\alpha} h(t)$ we have (2.10) according as $I(\gamma) < \infty$ $(= \infty)$. By (2.23) we have $I(\gamma) < \infty$ $(= \infty)$ according as

$$\int_1^\infty \frac{1}{t[h(t)]^\alpha} \, dt < \infty \quad (= \infty)$$

so the theorem is proved by scaling the function $h(t)$.

remark 3. If $\alpha = 2$ in Theorem 4, then the integral tests of [19] are more general. However, Theorem 3, along with the exponential decay of the tail of a Gaussian random variable as obtained in [14], easily gives

$$\overline{\lim_{t \to \infty}} \frac{|X(t)|}{t^{\frac{1}{2}} \sqrt{\log \log t}} = \Gamma \qquad \text{w.p. } 1$$

where Γ is a positive finite constant. This is well known so it will not be repeated here.

3. Some examples of the rate of escape with respect to the ℓ_∞ norm.

For our first set of examples we let

$$X(t) = \sum_{k=1}^{\infty} \frac{B_k(t) e_k}{a_k} \qquad (t \geq 0) \tag{3.1}$$

where $\{B_k(t) : k \geq 1, t \geq 0\}$ is a sequence of mutually independent sample continuous one dimensional Brownian motions with $EB_k(t) = 0$ and $EB_k^2(t) = t$ for all $t \geq 0$, $k \geq 1$. Here e_k is the kth unit coordinate vector in $\mathbb{R}^\infty$ for $k \geq 1$, and $\{a_k\}$ is a strictly positive increasing sequence.

If $\mathbb{R}^\infty$ is given the product topology, then it is easy to see that $\{X(t) : t \geq 0\}$ is sample continuous and for $\{x_k\}$ in $\mathbb{R}^\infty$ we define

$$|\{x_k\}|_\infty = \sup_{k \geq 1} |x_k| .$$

Then it is easy to show that:

(i) If $P(|X(t)|_\infty < \infty) > 0$ for any $t > 0$, then

$$(\log k)^{\frac{1}{2}} = O(a_k) \quad \text{as} \quad k \to \infty . \tag{3.2}$$

(ii) If $P(X(t) \in c_0) > 0$ for any $t > 0$, then

$$(\log k)^{\frac{1}{2}} = o(a_k) \quad \text{as} \quad k \to \infty .$$

Here c_0 is the subspace of $\mathbb{R}^\infty$ consisting of sequences which converge to zero. Furthermore, the converses of (3.2 - i) and (3.2 - ii) hold with each of the probabilities identically one. However, the main point of the converse of (3.2 - ii) is that if $(\log k)^{\frac{1}{2}} = o(a_k)$ as $k \to \infty$, then $\{X(t) : t \geq 0\}$

can actually be taken as a sample continuous Brownian motion in the separable Banach space $(c_0, |\cdot|_\infty)$, and this is what we assume throughout.

Theorem 5. If $a_k = k^p L(k)$ is a non-decreasing sequence with $0 < p < \infty$ and $L(x)$ is a strictly positive slowly varying function such that $L(1) = 1$, then with probability one

$$(3.3) \qquad \liminf_{t \to \infty} \left(\frac{\log \log t}{L^*(\log \log t)} \right)^p \frac{|X(t)|_\infty}{\sqrt{t}} > 0$$

where L^* is any slowly varying function such that for some $\rho > 0$

$$(3.4) \qquad L^*(x^p L(x)) \sim \frac{\rho}{L^{1/p}(x)} \quad \text{as } x \to \infty, \text{ and}$$

$$(3.5) \qquad \liminf_{x \to \infty} L^*(x^p/A(L^*(x))^p) \cdot \frac{1}{L^*(x)} = \lambda(p) > 0$$

for all $A > 0$. In addition, if for all sufficiently large n there exists $c_1, c_2 > 0$ satisfying

$$\sum_{k=1}^{n} \log L(k) \geq (n - c_1) \log L(n) - c_2 n,$$

and L^* is a slowly varying function such that for each $A > 0$

$$(3.6) \quad 0 < c_3 = \varliminf_n L\left(\frac{\log n}{A^{1/p}}\right) \{L^*(\log n)\}^p \leq \varlimsup_n L\left(\frac{\log n}{A^{1/p}}\right) L^*(\log n)^p = c_4 < \infty,$$

then we also have

$$(3.7) \qquad \liminf_{t \to \infty} \left(\frac{\log \log t}{L^*(\log \log t)} \right)^p \frac{|X(t)|_\infty}{\sqrt{t}} < \infty.$$

remark 1. Assuming all the conditions required of L and L^* in the theorem we have by combining (3.3) and (3.7), and a standard zero-one argument, that with probability one

$$(3.8) \qquad \liminf_{t \to \infty} \left(\frac{\log \log t}{L^*(\log \log t)} \right)^p \frac{|X(t)|_\infty}{\sqrt{t}} = c$$

where $0 < c < \infty$. Thus infinite dimensional Brownian motions sometime have a natural rate of escape. This is in contrast to the Dvoretsky-Erdős result in finite dimensions which only guarantees that if $3 \leq d < \infty$, then with probability one

$$(3.9) \qquad \liminf_{t \to \infty} \frac{|X(t)|_\infty}{t^{\frac{1}{2}} h(t)} = 0 \quad (\text{or } \infty)$$

according as

$$(3.10) \qquad \sum_k h^{d-2}(2^k) = \infty \quad (\text{or} < \infty).$$

Also see remark 2 of section 2 for some additional information.

remark 2. It is easy to see that the normalization $L(1) = 1$ presents no loss of generality. Furthermore, if $L(x) \equiv 1$, then $L^*(x) = 1$ satisfies all the conditions of the theorem. If $L(x) = (\log x)^q$ for $x \geq e$ and one elsewhere where $q \in \mathbb{R}^1$, then for n sufficiently large

$$\sum_{k=1}^{n} \log L(k) \geq (n-1)\log L(n) - |q|n, \quad \text{and}$$

$$L^*(x) = \frac{1}{(\log x)^{q/p}}$$

satisfies all the conditions of the theorem with $\rho = p^{-q/p}$, and $\lambda(p) = p^{-q/p}$. Other examples are also easily obtained.

remark 3. Since L is slowly varying the existence of a function L^* satisfying (3.4) is always guaranteed by [20, p. 21].

Proof of Theorem 5. The proof will be a simple combination of the following two lemmas.

Lemma 1. If $a_k = k^p L(k)$ is a non-decreasing sequence with $0 < p < \infty$ and $L(x)$ is a strictly positive continuous slowly varying function such that $L(1) = 1$, then with probability one

$$\lim_{t \to \infty} \left(\frac{\log\log t}{L^*(\log\log t)}\right)^p \frac{|X(t)|_\infty}{\sqrt{t}} > 0 \tag{3.11}$$

provided L^* is any slowly varying function satisfying (3.4) and (3.5).

remark 4. The sequence $\{a_k\}$ is eventually strictly increasing since $0 < p < \infty$ and $L(x)$ is a slowly varying strictly positive function.

Proof of Lemma 1. To establish the lemma we apply Theorem 2 with $b = e$ and $\alpha = 2$. That is, we will show there exists $A > 0$ sufficiently small so that

$$\sum_n P(|X(1)|_\infty \leq Ah(e^n))(h(e^n))^{-2} < \infty \tag{3.12}$$

with

$$h(t) = \frac{1}{(\log\log t / L^*(\log\log t))^p} \tag{3.13}$$

and L^* is as in (3.4) and (3.5).

Using the inequality $1 - x \leq e^{-x}$ for $x \geq 0$ we have

$$P(|X(1)|_\infty \leq Ah(e^n)) \leq \exp\{-\sum_{k=1}^{\infty} P(|\eta| \geq Ah(e^n)a(k)\} \tag{3.14}$$

where $a(k) = a_k$ and η is a mean zero Gaussian random variable with variance one. Now

$$J_n \equiv \sum_{k=1}^{\infty} P(|\eta| \geq Ah(e^n)a(k)) \geq \sum_{k=r}^{\infty} P(|\eta| \geq a(k)Ah(e^n)) \tag{3.15}$$

where $r > 1$ is chosen so that $a(r) \geq \Gamma > 1$ and $a^{-1}([\Gamma, \infty)) > 1$. Of course, $a^{-1}(x)$ exists for large x by remark 4. Hence for all n sufficiently large and all $0 < A \leq 1$

$$J_n \geq \sum_{k=r}^{\infty} P\left(a^{-1}\left(\frac{|\eta|}{Ah(e^n)}\right) > k;\ \frac{|\eta|}{Ah(e^n)} > \Gamma\right)$$

$$(3.16) \qquad \geq E\left(a^{-1}\left(\frac{|\eta|}{Ah(e^n)}\right) 1\left\{\frac{|\eta|}{Ah(e^n)}\right\} > \Gamma\right)$$

$$\geq \frac{1}{2\rho} E\left(\frac{|\eta|^{1/p}}{(Ah(e^n))^{1/p}} \; L^*\left(\frac{|\eta|}{Ah(e^n)}\right) 1\left\{\frac{|\eta|}{Ah(e^n)}\right\} > \Gamma\right)$$

since $a^{-1}(x) \sim \frac{x^{1/p}}{\rho} L^*(x)$. That is, $a^{-1}(x) \geq 1$ on $[\Gamma, \infty)$ and by [20, p. 21] $a^{-1}(x) \sim x^{1/p}\tilde{L}(x)$ where $\tilde{L}(x)$ is slowly varying. Hence for large x

$$(3.17) \qquad \begin{aligned} x &= a^{-1}(a(x)) \\ &= a^{-1}(x^p L(x)) \sim xL^{1/p}(x)\tilde{L}(x^p L(x)) \end{aligned}$$

and hence as $x \to \infty$

$$(3.18) \qquad \tilde{L}(x^p L(x)) \sim \frac{1}{L^{1/p}(x)} .$$

Thus (3.4) and (3.18) imply

$$a^{-1}(x) \sim \frac{x^{1/p} L^*(x)}{\rho} .$$

Now the dominated convergence theorem and L^* being slowly varying implies

$$(3.19) \qquad \frac{1}{L^*(1/Ah(e^n))} E\left\{|\eta|^{1/p} L^*\left(\frac{|\eta|}{Ah(e^n)}\right) 1_{\{|\eta|/Ah(e^n) > \Gamma\}}\right\} \xrightarrow{n\to\infty} E|\eta|^{1/p} .$$

Hence for n large we have that

$$(3.20) \qquad J_n \geq L^*(1/Ah(e^n)) \frac{E|\eta|^{1/p}}{4\rho(Ah(e^n))^{1/p}} .$$

Combining (3.14), (3.15), and (3.20) we have for n large and $0 < A \leq 1$ that

$$(3.21) \qquad P(|X(1)|_\infty \leq Ah(e^n)) \leq \exp\{-L^*\left(\frac{1}{Ah(e^n)}\right) E|\eta|^{1/p}/4\rho(Ah(e^n))^{1/p}\}$$

$$= \exp\{-E|\eta|^{1/p} \log n \, L^*\left(\frac{(\log n)^p}{A(L^*(\log n))^p}\right) /4\rho A^{1/p} L^*(\log n)\} .$$

Now fix A, $0 < A \leq 1$, such that

$$(3.22) \qquad \frac{E|\eta|^{1/p}}{8A^{1/p}\lambda(p)} > 2\rho$$

when $\lambda(p)$ is as in (3.5). Combining (3.21), (3.22) and using (3.5) we have for n sufficiently large that

$$(3.23) \qquad P(|X(1)|_\infty \leq Ah(e^n)) \leq \exp\{-2\log n\} .$$

Hence from (3.13) and (3.23) there exists A, $0 < A \leq 1$, such that (3.12) holds since L^* slowly varying implies

$$x^{-\varepsilon} \le L^*(x) \le x^{\varepsilon}$$

for any $\varepsilon > 0$ and all sufficiently large x. Hence the lemma is proved.

Lemma 2. If $a_k = k^p L(k)$ is a non-decreasing sequence with $0 < p < \infty$ and $L(x)$ is a strictly positive slowly varying function such that $L(1) = 1$ and for all sufficiently large n there exists $c_1, c_2 > 0$ such that

$$\sum_{k=1}^{n} \log L(k) \ge (n - c_1)\log L(n) - c_2 n,$$

then with probability one

$$(3.24) \qquad \liminf_{t \to \infty} \left(\frac{\log\log t}{L^*(\log\log t)}\right)^p \frac{|X(t)|_\infty}{\sqrt{t}} < \infty$$

provided L^* is a slowly varying function such that for each $A > 0$

$$(3.25) \quad 0 < c_3 = \underline{\lim}_{n}\, L(\log n/A^{1/p})\{L^*(\log n)\}^p \le \overline{\lim}_{n}\, L\left(\frac{\log n}{A^{1/p}}\right)(L^*(\log n))^p = c_4 < \infty .$$

Proof. To establish the lemma we again apply Theorem 2 with $b = e$ and $\alpha = 2$. That is, we will show there exists $A > 0$ sufficiently large so that

$$(3.26) \qquad \sum_n P(|X(1)|_\infty \le Ah(e^n))(h(e^n))^{-2} = \infty$$

with $h(t)$ defined as in (3.13).

To verify (3.26) holds for A sufficiently large we proceed as follows. First define $r_n = \max(1, \left[\frac{\log n}{A^{1/p}}\right])$ where $[\cdot]$ equals the greatest integer function and A is to be specified later. Then we define

$$(3.27) \qquad B_n = P\left(\sup_{k \le r_n} \frac{|B_k(1)|}{a(k)} \le \frac{A(L^*(\log n))^p}{(\log n)^p}\right)$$

and

$$(3.28) \qquad C_n = P\left(\sup_{k > r_n} \frac{|B_k(1)|}{a(k)} \le \frac{A(L^*(\log n))^p}{(\log n)^p}\right).$$

Letting $\eta = N(0,1)$ we have $c_5 > 0$ such that

$$\frac{1}{C_n} = \prod_{k > r_n}\left\{\frac{1}{1 - P\left(|\eta| > Ak^p L(k)\ \frac{(L^*(\log n))^p}{(\log n)^p}\right)}\right\}$$

$$\le \prod_{k > r_n}\left\{1 + c_5 P\left(|\eta| > Ak^p L(k)\ \frac{(L^*(\log n))^p}{(\log n)^p}\right)\right\}$$

since for all n sufficiently large and $k \ge r_n$ we have by (3.25) that

$$Ak^p \frac{L(k)}{(\log n)^p} (L^*(\log n))^p \geq \frac{c_3}{2} .$$

Hence for n sufficiently large (3.25) again implies

(3.29)
$$\begin{aligned}
\frac{1}{C_n} &\leq \exp\left\{c_5 \sum_{k>r_n} P(|\eta| > \frac{Ak^p}{(\log n)^p})\right\} \\
&\leq \exp\left\{c_5 \int_0^\infty P(|\eta|^{1/p} \frac{\log n}{A^{1/p}} > x)\, dx\right\} \\
&= \exp\left\{c_5 \frac{\log n}{A^{1/p}} E|\eta|^{1/p}\right\} \\
&= n^{\frac{m}{A^{1/p}}}
\end{aligned}$$

where $m = c_5 E|\eta|^{1/p}$.

Let $f(u) = \frac{1}{\sqrt{2\pi}} e^{-u^2/2}$ denote the density of η. Then for given A and large n

(3.30)
$$\begin{aligned}
B_n &= \prod_{k \leq r_n} \left(P(|\eta| \leq Ak^p L(k) \frac{(L^*(\log n))^p}{(\log n)^p}\right) \\
&\geq \prod_{k \leq r_n} \left\{\frac{2Ak^p L(k)}{(\log n)^p} (L^*(\log n))^p f(2c_4)\right\} \\
&\geq \left(\frac{2AL^*(\log n)^p f(2c_4)}{(\log n)^p}\right)^{r_n} (r_n!)^p \left\{\prod_{k \leq r_n} L(k)\right\}
\end{aligned}$$

Furthermore, since there exists constants $c_1, c_2 > 0$ such that

(3.31)
$$\sum_{k=1}^{n} \log L(k) \geq (n - c_1) \log L(n) - c_2 n$$

we have for all sufficiently large n that

(3.32)
$$\begin{aligned}
\prod_{k \leq r_n} L(k) &= \exp\left\{\sum_{k=1}^{r_n} \log L(k)\right\} \\
&\geq \frac{L(r_n)^{r_n - c_1}}{e^{c_2 r_n}} .
\end{aligned}$$

By combining (3.30), (3.31) and (3.32) we thus have for all sufficiently large n that

$$B_n \geq \left(\frac{2A(L^*(\log n))^p f(2c_4) L(r_n)}{(\log n)^p e^{c_2}}\right)^{r_n} \frac{(r_n!)^p}{L(r_n)^{c_1}}$$

(3.33)
$$\geq \left(\frac{f(2c_4)\, 2A\, c_3/2}{e^{c_2}(\log n)^p}\right)^{r_n} \frac{\left(r_n^{\,r_n+1/2}\right)^p e^{-r_n p} (\sqrt{2\pi})^p}{2L(r_n)^{c_1}}$$

$$\geq c_6 \frac{(\log n)^{p/2}}{A\,L\left(\left[\frac{\log n}{A^{1/p}}\right]\right)^{c_1}} \left(\frac{f(2c_4)c_3 e^{-p}}{e^{c_2}}\right)^{\frac{\log n}{A^{1/p}}}$$

where $c_6 > 0$ is an absolute constant independent of $A \geq 1$ and $n \geq 3$.

Hence for all sufficiently large n

$$P(|X(1)|_\infty \leq Ah(e^n)) = B_n C_n$$

(3.34)

$$\geq \frac{c_6(\log n)^{p/2}}{AL\left(\left[\frac{\log n}{A^{1/p}}\right]\right)^{c_1}} \; n^{-\frac{m}{A^{1/p}} + \frac{\log \Gamma}{A^{1/p}}}$$

where $\Gamma = e^{-p} f(2c_4)c_3/e^{c_2}$ and $m = c_5 E|\eta|^{1/p}$. Thus choosing A sufficiently large so that

$$-1 < (-m + \log \Gamma)/A^{1/p}$$

we have from (3.34) that (3.26) holds and the lemma is proved.

Now let

(3.35)
$$X(t) = \sum_{k=1}^{\infty} \frac{\theta_k(t) e_k}{a_k} \qquad (t \geq 0)$$

where $\{\theta_k(t) : k \geq 1,\ t \geq 0\}$ is a sequence of mutually independent symmetric stable processes of index $\alpha < 2$ with sample paths in $D[0,\infty)$ and having stationary independent increments with $\theta_k(0) = 0$. Again e_k, $|\{x_k\}|_\infty$, and $\{a_k\}$ are as described following (3.1).

Then it is easy to show that

(3.36) $\quad P(|X(t)|_\infty < \infty) > 0$ for some $t > 0$ iff $\sum_k a_k^{-\alpha} < \infty$,

and, of course, since $\{a_k\}$ is increasing the left-hand side of (3.36) therefore implies $k^{1/\alpha} = 0(a_k)$. Further, since $\{|X(t)|_\infty < \infty\}$ is a tail event for the sequence $\{\theta_k(t) : k \geq 1\}$ we have $P(|X(t)|_\infty < \infty) = 1$ for some t

(and hence all t by scaling) iff $\sum_k a_k^{-\alpha} < \infty$. Hence it is easy to see that if $a_k = k^p L(k)$ where L is a slowly varying function and $p > \frac{1}{\alpha}$, then $P(X(t) \in c_0) = 1$ for all $t \geq 0$. In particular, for $p > \frac{1}{\alpha}$ we thus have $X(t)$ a symmetric stable process with stationary independent increments in the Banach space $(c_0, |\cdot|_\infty)$, $X(0) = 0$, and it is easy to prove that $X(t)$ has a separable version with right continuous sample paths which have left hand limits. This we assume throughout and hence we can prove

Theorem 6. If $X = \{X(t) : t \geq 0\}$ is a separable symmetric stable process of index $\alpha < 2$ with stationary independent increments satisfying (3.35) where $a_k = k^p L(k)$, $p > \frac{1}{\alpha}$, is a non-decreasing sequence and $L(x)$ is a strictly positive slowly varying function such that $L(1) = 1$, then probability one

$$\varliminf_{t\to\infty} \left(\frac{\log\log t}{L^*(\log\log t)}\right)^p \frac{|X(t)|_\infty}{t^{1/\alpha}} > 0 \tag{3.37}$$

where L^* is any slowly varying function such that for some $\rho > 0$ (3.4) and (3.5) hold for all $A > 0$. In addition, if for all sufficiently large n there exists $c_1, c_2 > 0$ satisfying $\sum_{k=1}^{n} \log L(k) \geq (n-c_1)\log L(n) - c_2 n$, and L^* is a slowly varying function such that for each $A > 0$ (3.6) holds, then we also have

$$\varliminf_{t\to\infty} \left(\frac{\log\log t}{L^*(\log\log t)}\right)^p \frac{|X(t)|_\infty}{t^{1/\alpha}} < \infty . \tag{3.38}$$

The proof of Theorem 6 is exactly as the proof of Theorem 5 except for a couple of minor changes. For example, for Theorem 6 η is assumed to be a symmetric stable random variable of index α and since $p > \frac{1}{\alpha}$ we have $E(|\eta|^{1/p}) < \infty$. Another minor change appears in the proof of Lemma 2 where $f(u)$ is now assumed to be the density of the symmetric stable random variable η, and in (3.30) we replace $f(2c_4)$ by $c_7 = \inf_{|u| \leq 2c_4} f(u)$. Of course, $c_7 > 0$ since $f(u)$ is a positive continuous function on $\mathbb{R}^1$ in the stable case, so the proof goes as before.

4. An application to Kiefer Processes and the empirical process.

Let $\{\eta_n : n \geq 1\}$ be a sequence of random variables uniformly distributed on $[0, 1]$ and let $F_N(s)$, $0 \leq s \leq 1$, be its empirical distribution function at stage N. The empirical process of $\{\eta_n : n \geq 1\}$ is defined by

$$R(s, t) = t(F_{[t]}(s) - s) \quad 0 \leq s \leq 1, \quad t \geq 0 \tag{4.1}$$

where $[t]$ denotes the greatest integer not exceeding t. Modifying the terminology of [2] slightly we say a separable Gaussian process $K(s, t)$ on $[0, 1] \times [0, \infty)$ is a Kiefer process if $K(0, t) = K(s, 0) = 0$ for all $0 \leq s \leq 1$, $t \geq 0$, and

$$(4.2)\qquad \begin{aligned} EK(s,t) &= 0 \\ E\,K(s,t)K(s',t') &= \min(t,t')\,\Gamma(s,s') \end{aligned}$$

where $\Gamma(s,s')$ is the covariance function of a separable Gaussian process $\{G(s): 0 \le s \le 1\}$ with $G(0) = 0$.

In case $\Gamma(s,s') = \min(s,s')$, then the process $\{K(s,t): 0 \le s \le 1, t \ge 0\}$ is often called a Brownian sheet as $G(s)$ is standard Brownian motion on $[0,1]$, and if $\Gamma(s,s') = s(1-s')$ for $0 \le s \le s' \le 1$ then we refer to it as the tied down Brownian sheet. Of course, in the latter case $G(s)$ is the Brownian bridge on $[0,1]$. It is well known that in each of these situations $\{K(s,t): 0 \le s \le 1, 0 \le t\}$ has a version which has continuous sample paths. Furthermore if $X = \{X(t): 0 \le t < \infty\}$ is defined by

$$(4.3)\qquad X(t)(\cdot) = K(\cdot, t) \qquad (t \ge 0),$$

then from [18] we easily have that $\{X(t)\}$ is a sample continuous Brownian motion in the Banach space $C[0,1]$ determined by the mean zero Gaussian measure $\mu = \mathcal{L}(X(1))$.

Thus we can apply Theorem 2 to obtain

<u>Theorem 7.</u> If $\{K(s,t): 0 \le s \le 1, t \ge 0\}$ is a sample continuous Brownian sheet or a tied down Brownian sheet, then with probability one

$$(4.4)\qquad \liminf_{t\to\infty} \sqrt{\log\log t}\ \sup_{0\le s\le 1} \frac{|K(s,t)|}{\sqrt{t}} = \frac{\pi}{\sqrt{8}}$$

and

$$(4.5)\qquad \liminf_{t\to\infty} \sqrt{\log\log t}\ \left(\int_0^1 \frac{K^2(s,t)\,ds}{t}\right)^{\frac{1}{2}} = \frac{1}{\sqrt{8}}\ .$$

Proof. If $\{K(s,t)\}$ is the Brownian sheet and $\{X(t)\}$ is defined by (4.3), then $\mathcal{L}(X(1)) = \mu$ where μ is Wiener measure. By [7, p. 206]

$$(4.6)\qquad P(\|X(1)\|_\infty \le t) \sim \frac{4}{\pi}\ e^{-\pi^2/8t^2}$$

as $t \downarrow 0$, and by [6]

$$(4.7)\qquad P(\|X(1)\|_2 \le t) \sim \sqrt{2/\pi}\ \ e^{-1/8t^2}\ \ 2\sqrt{2t}$$

as $t \downarrow 0$. Here, of course, $\|\cdot\|_\infty$ denotes the sup-norm on $C[0,1]$ and $\|\cdot\|_2$ denotes the L^2-norm on $C[0,1]$.

To verify (4.4) we let $h(t) = \dfrac{1}{\sqrt{\frac{8}{\pi^2}\log\log t}}$. Then Theorem 2 and (4.6) implies

$$(4.8)\qquad \liminf_{t\to\infty} \frac{\|X(t)\|_\infty}{t^{\frac{1}{2}}h(t)(1-\varepsilon)} \ge 1$$

for any ε, $0 < \varepsilon < 1$, and similarly

$$(4.9) \qquad \liminf_{t \to \infty} \frac{\|X(t)\|_\infty}{t^{\frac{1}{2}} h(t)(1+\varepsilon)} \leq 1 .$$

Thus (4.4) holds as claimed.

To verify (4.5) we apply (4.7) and Theorem 2 with

$$h(t) = \frac{1}{\sqrt{8 \log \log t}}$$

to obtain the analogues of (4.8) and (4.9) for the L^2-norm. Hence (4.5) holds.

To prove (4.4) and (4.5) for the tied down Brownian sheet we proceed in exactly the same manner. In this situation $\mathcal{L}(X(1))$ is the measure induced on $C[0,1]$ by the Brownian bridge and hence by [3, p. 85] and [1, p. 302] we have

$$(4.10) \qquad P(\|X(1)\|_\infty \leq t) = \sum_{j=-\infty}^{\infty} (-1)^j e^{-2j^2 t^2}$$

and

$$(4.11) \qquad P(\|X(1)\|_2 \leq t) = \frac{1}{\pi t} \sum_{j=-\infty}^{\infty} \frac{\Gamma(j+\frac{1}{2})}{\Gamma(j) j!} (4j+1)^{\frac{1}{2}} \exp\left\{\frac{-(4j+1)^2}{16t^2}\right\} K_{\frac{1}{4}}\left\{\frac{(4j+1)^2}{16t^2}\right\}$$

where $K_{\frac{1}{4}}(x)$ is the standard Bessel function.

Now using the Poisson summation formula [13, p. 630] as employed in [9, p. 35-36] we have

$$(4.12) \qquad \sum_{j=-\infty}^{\infty} (-1)^j e^{-2j^2 t^2} = \frac{\sqrt{2\pi}}{t} \sum_{k=0}^{\infty} \exp\left\{\frac{-(2k+1)^2 \pi^2}{8t^2}\right\}$$

$$\sim \frac{\sqrt{2\pi}}{t} \exp\left\{-\frac{\pi^2}{8t^2}\right\}$$

as t decreases to zero. Hence (4.4) holds by applying Theorem 2 as above.

Since $K_{1/4}(x) \sim \frac{e^{-x}}{\sqrt{2x/\pi}}$ as x tends to infinity [16, p. 146] the first term of the series in (4.11) dominates as $t \downarrow 0$, so we obtain

$$(4.13) \qquad P(\|X(1)\|_2 \leq t) \sim 2\sqrt{2} \exp\left\{-\frac{1}{8t^2}\right\} .$$

Hence (4.5) holds by applying Theorem 2 as above and the theorem is proved.

As an application of Theorem 7 and the approximation result in [17] or [2] we thus have

<u>Theorem 8.</u> Let $\{\eta_n : n \geq 1\}$ be a sequence of independent random variables uniformly distributed on $[0,1]$ and let $\{R(s,t) : 0 \leq s \leq 1, t \geq 0\}$ denote the empirical process of (4.1). Then, with probability one

(4.14) $$\liminf_{t \to \infty} \sqrt{\log \log t} \sup_{0 \le s \le 1} \frac{|R(s,t)|}{\sqrt{t}} = \frac{1}{\sqrt{8}}$$

and

(4.15) $$\liminf_{t \to \infty} \sqrt{\log \log t} \left(\int_0^1 \frac{R^2(s,t)ds}{t} \right)^{\frac{1}{2}} = \frac{1}{\sqrt{8}} .$$

Proof. Using the approximation result of [2] or [17] we can have a sample continuous tied down Brownian sheet $\{K(s,t) : 0 \le s \le 1, t \ge 0\}$ such that with probability one

(4.16) $$\sup_{0 \le t \le T} \sup_{0 \le s \le 1} \frac{|R(s,t) - K(s,t)|}{T^{\frac{1}{2}}(\log T)^{-\lambda}} = O(1)$$

for some $\lambda > 0$. Hence with probability one

(4.17) $$\lim_{t \to \infty} \sup_{0 \le s \le 1} \frac{|R(s,t) - K(s,t)|}{\sqrt{t}} \sqrt{\log \log t} = 0$$

and (4.14) and (4.15) follow immediately from (4.4) and (4.5) for a tied down Brownian sheet.

remark By applying the results of [2] and the ideas of Theorem 7 it is easy to see that Theorem 8 can be extended to many strictly stationary sequences of strong mixing random variables provided one can obtain the behavior of the probability $P(\|G(s)\| < \varepsilon)$ as $\varepsilon \downarrow 0$ when G has covariance $\Gamma(s, s')$ and $\|\cdot\|$ is some suitable norm. However, these estimates appear to be non-trivial, so much remains to be done in this regard.

Bibliography

[1] Anderson, T. W. and Darling, D.A. (1952), Asymptotic theory of certain "goodness of fit" criteria based on stochastic processes, Ann. Math. Stat. Vol. 23, pp. 193-212.

[2] Berkes, I. and Philipp, W. (1977), An almost sure invariance principle for the empirical distribution function of mixing random variables, Z. Wahr. Verw. Gebiete, Vol. 41, pp. 115-137.

[3] Billingsley, P. (1968), Convergence of Probability Measures, J. Wiley & Sons, New York.

[4] Bojanic, R. and Seneta, E. (1971), Slowly varying functions and asymptotic relations, J. Math. Anal. Appl., Vol. 34, pp. 302-315.

[5] Breiman, L. (1968), Probability, Addison Wesley, Publ. Co., Reading, Massachusetts.

[6] Cameron, R. H. and Martin, W. T. (1944), The Wiener measure of Hilbert neighborhoods in the space of real continuous functions, J. of Math. and Physics, Vol. 23, pp. 195-209.

[7] Chung, K. L. (1948), On the maximum partial sums of sequences of independent random variables, Trans. Amer. Math. Soc., Vol. 64, pp. 205-233.

[8] De Acosta, Alejandro (1975), Stable Measures and Seminorms, Annals of Prob., Vol. 3, pp. 865-875.

[9] Dudley, R. M., Hoffman-Jørgensen, J. and Shepp, L. A. (1977), On the lower tail of Gaussian seminorms, Aarhus Universitet Preprint series.

[10] Dudley, R. M. and Kanter, M. (1974), Zero-One laws for Stable Measures, Proc. American Math. Soc., Vol. 25, pp. 245-252.

[11] Dvoretsky, A. and Erdős, P. (1951), Some problems on random walk in space, Proc. Second Berkeley Sym. Math., Statistics, and Prob., pp. 353-367.

[12] Erickson, K. B. (1978), Rates of escape of infinite dimensional Brownian motion, Submitted for publication.

[13] Feller, W. (1970), An introduction to probability theory and its applications, Vol. 2, second edition, J. Wiley & Sons, New York.

[14] Fernique, X. (1970), Integrabilité des vecteurs Gaussiens, C. R. Acad. Sci. Paris Ser. A, Vol. 270, pp. 1698-1699.

[15] Gihman, I. I. and Skorohod, A. V. (1974), The Theory of Stochastic Processes I, Springer-Verlag, Berlin.

[16] Hildebrand, F. B., Advanced Calculus for Applications, second edition, Prentice Hall Inc., Englewood Cliffs, New Jersey.

[17] Kiefer, J. (1972), Skorohod embedding of multivariate random variables and the sample D. F., Z. Wahr. Verw. Gebiete, Vol. 24, pp. 1-35.

[18] Kuelbs, J. (1973), The invariance principle for Banach space valued random variables, J. Multivariate Analysis, Vol. 3, pp. 161-172.

[19] ________________ (1975), Sample path behavior for Brownian motion in Banach spaces, Ann. of Prob., Vol. 3, pp. 247-261.

[20] Seneta, E. (1976), Regularly varying functions, Lecture Notes in Mathematics, Vol. 508, Springer-Verlag, Berlin.

[21] Takeuchi, J. (1964), On the sample paths of the symmetric stable processes in spaces, J. Math. Soc. Japan, Vol. 16, pp. 109-127.

James Kuelbs
Department of Mathematics
University of Wisconsin
Madison, Wisconsin 53706

ALMOST SURE INVARIANCE PRINCIPLES FOR SUMS OF B-VALUED RANDOM VARIABLES

Walter Philipp
Department of Mathematics
University of Illinois
Urbana, Illinois 61801/USA

1. Introduction

The main purpose of this paper is to give necessary and sufficient conditions for the almost sure approximation of the t-th partial sum of independent identically distributed random variables with values in a separable Banach space and having finite second moments by a suitable Brownian motion $\{X(t), t \geq 0\}$ with an error term $o((t \log\log t)^{\frac{1}{2}})$. This is a synopsis of the following theorem.

Theorem 1. Let $\{x_\nu, \nu \geq 1\}$ be a sequence of independent identically distributed random variables centered at expectations, with finite second moments and assuming values in a separable Banach space B. Then the following two statements are equivalent.

(a) There exists a probability space and a Brownian motion $\{X(t), t \geq 0\}$ defined on it with covariance structure

$$T(f,g) = E\{f(x_1)g(x_1)\} \qquad f,g \in B^*$$

and a sequence $\{y_\nu, \nu \geq 1\}$ of random variables having the same distribution as $\{x_\nu, \nu \geq 1\}$ such that with probability 1

$$\| \textstyle\sum_{\nu \leq t} y_\nu - X(t) \| = o((t \log\log t)^{\frac{1}{2}}) \qquad t \to \infty. \tag{1.1}$$

(b) x_1 is pregaussian and any of the following three conditions holds.

(i) $(n \log\log n)^{-\frac{1}{2}} \sum_{\nu \leq n} x_\nu \to 0$ in probability

(ii) $(n \log\log n)^{-\frac{1}{2}} E\| \sum_{\nu \leq n} x_\nu \| \to 0$

Supported in part by a grant from the National Science Foundation. AMS (1970) Subject classification, 60B99, 60F15.

(iii) $\{x_\nu, \nu \geq 1\}$ satisfies the compact law of the iterated logarithm with limit set K, the unit ball of the Hilbert space $H_{L(x_1)}$.

Recall that a random variable x is pregaussian if there is a Gaussian distribution on B with the same covariance structure as x. Since each $\mathbb{R}^d$-valued random variable with finite variance is pregaussian we have the following corollary.

Corollary 1. Let $\{x_\nu, \nu \geq 1\}$ be a sequence of independent identically distributed random variables with values in $\mathbb{R}^d$, centered at expectation and with finite variances. Then without changing its distribution we can redefine the sequence $\{x_\nu, \nu \geq 1\}$ on a new probability space on which there exists Brownian motion $\{X(t), t \geq 0\}$ with the same covariance matrix as x_1 such that

$$\sum_{\nu \leq t} x_\nu - X(t) = o((t \log \log t)^{\frac{1}{2}}) \qquad \text{a.s.}$$

Theorem 1 is the Banach space analogue of Strassen's (1964) Theorem 2. Under the assumption of the finiteness of only the second moments the error term in (1.1) cannot be improved as was shown by Major (1976a) for real-valued random variables. Exactly the same result as Theorem 1, but with the additional assumption of the finiteness of the $(2 + \delta)$-th moments was recently proved by Kuelbs and Philipp (1977) by an argument similar in spirit but differing rather markedly in the details. Whether in this case the error term could be improved without some additional assumptions on the structure of B is doubtful. In $\mathbb{R}^d$, however, Theorem 2 below yields an error term which is almost best possible.

The following corollary is an immediate consequence of Theorem 1 and a theorem of Pisier (1975) which says that under the hypotheses of the corollary the compact law of the iterated logarithm holds. Moreover, it is easy to see that under the hypotheses of the corollary x_1 is pregaussian.

Corollary 2. Let $\{x_\nu, \nu \geq 1\}$ be a sequence of independent identically distributed B-valued random variables centered at expectations and with finite second moments. Suppose that $\{x_\nu, \nu \geq 1\}$ satisfies the central limit theorem. Then statement (a) of Theorem 1 holds.

As a matter of fact the proof of Theorem 1 yields the following corollary.

Corollary 3. Let $\{x_\nu, \nu \geq 1\}$ be a sequence of independent identically distributed B-valued random variables satisfying the central limit theorem with limiting Gaussian distribution λ and the compact law of the iterated logarithm. Then without changing its distribution we can redefine the sequence $\{x_\nu, \nu \geq 1\}$ on a new probability space on which there exists Brownian motion $\{X(t), t \geq 0\}$ with the same covariance structure as λ such that

$$\| \sum_{\nu \leq t} x_\nu - X(t) \| = o((t \log \log t)^{\frac{1}{2}}) \qquad \text{a.s.}$$

At the same conference in Oberwolfach Heinkel (1979) gave an improvement over Pisier's (1975) result by requiring only the finiteness of somewhat less than second moments. Combining Heinkel's theorem with Corollary 3 we obtain an almost sure invariance principle under the hypotheses of Heinkel's theorem.

In order for (1.1) to hold x_1 has to be pregaussian. Hence Theorem 1 does not contain Theorem 4.1 of Kuelbs (1977) which establishes the equivalence of conditions (i), (ii) and (iii) under the hypotheses of Theorem 1. But the following corollary to the proof of Theorem 1 does contain Theorem 4.1 of Kuelbs (1977).

Corollary 4. Let $\{x_\nu, \nu \geq 1\}$ be a sequence of independent identically distributed B-valued random variables centered at expectations and with finite second moments. Then the following statement is equivalent with each of the conditions (i), (ii) or (iii) in statement (b).
(iv) Given any $\alpha > 0$ we can redefine the sequence $\{x_\nu, \nu \geq 1\}$, without changing its distribution, on a new probability space on which there exists a Brownian motion $\{X_\alpha(t), t \geq 0\}$ with covariance structure $T_\alpha(f,g)$ such that

$$|T_\alpha(f,g) - T(f,g)| \leq \alpha \|f\| \, \|g\| \qquad f,g \in B^* \tag{1.2}$$

and

$$\limsup_{t\to\infty} (t \log \log t)^{-\frac{1}{2}} \| \sum_{\nu \leq t} x_\nu - X_\alpha(t) \| \leq \alpha \qquad \text{a.s.} \tag{1.3}$$

It might be worthwhile to remark that our construction will yield a Brownian motion with finite-dimensional range. A counterexample given in the paper by Kuelbs and Philipp (1977) shows that (iv) does not imply (1.1). This example shows that in the absence of the assumption that x_1 be pregaussian there simply is no universal

Brownian motion to approximate the partial sum process.

On the other hand if for each $0 < \alpha \leq \alpha_o$ there is a Brownian motion $\{X_\alpha(t), t \geq 0\}$ with covariance structure $T(f,g)$ (independent of α) such that (1.3) holds then (1.1) follows. As a matter of fact this is exactly the way the proof of Theorem 1 will proceed. (See section 3.4 below.)

As mentioned above for $\mathbb{R}^d$-valued random variables with finite $(2+\delta)$-th moments the following result gives an almost best possible error term in the approximation of the partial sums by Brownian motion.

<u>Theorem 2</u>. Let $\{x_\nu, \nu \geq 1\}$ be a sequence of independent identically distributed random variables with values in $\mathbb{R}^d$, centered at expectations and having finite $(2+\delta)$-th moments for some $0 < \delta \leq 1$. Then without changing its distribution we can redefine the sequence $\{x_\nu, \nu \geq 1\}$ on a new probability space on which there exists Brownian motion $\{X(t), t \geq 0\}$ with the same covariance matrix as x_1 such that

$$\sum_{\nu \leq t} x_\nu - X(t) \ll t^{1/(2+\delta)} \log t \qquad \text{a.s.}$$

For $d = 1$ Theorem 2 is due to Major (1976b) with error term $o(t^{1/(2+\delta)})$. This error term is best possible, as was also shown by Major (1976b). However, his method, based on the quantile transform, does not carry over to the case $d > 1$. Theorem 2 with a much weaker error term was proved by Berkes and Philipp (1979).

The proofs of all of these results as well as of the weak invariance principles (see Philipp (1978)) are based on the following approximation theorem which is a generalization of Theorem 2 of Berkes and Philipp (1979).

<u>Theorem 3</u>. Let $\{B_k, m_k, k \geq 1\}$ be a sequence of complete separable metric spaces. Let $\{X_k, k \geq 1\}$ be a sequence of random variables with values in B_k and let $\{L_k, k \geq 1\}$ be a sequence of σ-fields such that X_k is L_k-measurable. Suppose that for some sequence $\{\phi_k, k \geq 1\}$ of nonnegative numbers

$$|P(CD) - P(C)P(D)| \leq \phi_k P(C)$$

for all $k \geq 1$ and all $C \in \bigvee_{j<k} L_j$ and $D \in L_k$. Denote by F_k the distribution of X_k and let $\{G_k, k \geq 1\}$ be a sequence of distributions on B_k such that

(1.4) $$F_k(A) \leq G_k(A^{\rho_k}) + \sigma_k \quad \text{for all Borel sets} \quad A \subset B_k.$$

Here ρ_k and σ_k are nonnegative numbers and $A^{\varepsilon} = \cup_{x \in A}\{y: m_k(x,y) < \varepsilon\}$. Then without changing its distribution we can redefine the sequence $\{X_k, k \geq 1\}$ on a richer probability space on which there exists a sequence $\{Y_k, k \geq 1\}$ of independent random variables Y_k with distribution G_k such that for all $k \geq 1$

(1.5) $$P\{m_k(X_k,Y_k) \geq 2(\phi_k + \rho_k)\} \leq 2(\phi_k + \sigma_k).$$

Theorem 2 of Berkes and Philipp (1977) is obtained from Theorem 3 by setting $F_k = G_k$. The proof of Theorem 3 is given in the next section. The proofs of Theorem 1, Corollary 4, 3 and Theorem 2 will be given in sections 3, 4, 5 and 6 respectively.

2. Proof of Theorem 3.

The proof of Theorem 3 consists of an application of Theorem 2 of Berkes and Philipp (1979) and of the Strassen-Dudley theorem (see Strassen (1965) and Dudley (1968), Theorem 1) together with a rearrangement of some of the arguments given in the paper by Berkes and Philipp (1979). It is worth mentioning that if $\rho_k = \sigma_k$ then (1.4) obviously can be replaced by

$$\pi(F_k, G_k) \leq \rho_k$$

where π denotes the Prohorov distance on $\{B_k, m_k\}$.

For simplicity we write $|x - y|$ instead of $m_k(x,y)$. Using the separability of B_k we approximate each random variable X_k by a discrete random variable X_k^* such that X_k^* is L_k-measurable and that

(2.1) $$|X_k - X_k^*| \leq \tfrac{1}{4}\rho_k.$$

Let F_k^* be the distribution of X_k^*. By (2.1.7) in the proof of Theorem 2 of Berkes and Philipp (1979) there exists without loss of generality a sequence of independent random variables Z_k with distribution F_k^* such that

$$P\{|X_k^* - Z_k| \geq 2\phi_k\} \leq 2\phi_k. \tag{2.2}$$

Now let $\{\xi_k, k \geq 1\}$ be a sequence of independent random variables uniformly distributed over $[0,1]$ and let $M_k = \sigma(Z_k, \xi_k)$. Then $\{M_k, k \geq 1\}$ is a sequence of independent atomless σ-fields. We enrich the probability space (Ω, F, P) by adjoining the sequence $\{\xi_k, k \geq 1\}$ and denote this new space also by (Ω, F, P). We redefine the sequences $\{X_k, k \geq 1\}$, $\{X_k^*, k \geq 1\}$ and $\{Z_k, k \geq 1\}$ on this richer space without changing their joint distribution. By (2.1) and (1.4) we have for each Borel set A

$$F_k^*(A) = P(X_k^* \in A) \leq P(X_k \in A^{\frac{1}{4}\rho_k}) = F_k(A^{\frac{1}{4}\rho_k}) \leq G_k(A^{\frac{5}{4}\rho_k}) + \sigma_k.$$

Hence by the Strassen-Dudley theorem (see Dudley (1968), Theorem 1) there exists a probability measure Q_k on $B_k \times B_k$ with marginals F_k^* and G_k such that

$$Q_k\{(x,y): |x - y| \geq \tfrac{3}{2}\rho_k\} \leq \tfrac{3}{2}\sigma_k. \tag{2.3}$$

We apply Lemma 2.4 of Berkes and Philipp (1979) to the (enriched) space (Ω, M_k, P). The distribution F_k^* of Z_k is a marginal of Q_k. Hence there exists a random variable Y_k on (Ω, M_k, P) such that Z_k and Y_k have joint distribution Q_k. In particular we obtain from (2.3)

$$P\{|Z_k - Y_k| \geq \tfrac{3}{2}\rho_k\} \leq \tfrac{3}{2}\sigma_k. \tag{2.4}$$

Now Y_k has distribution G_k, the second marginal of Q_k. Since Y_k is M_k-measurable the random variables Y_k are independent. Finally (1.5) follows from (2.1), (2.2) and (2.4) since

$$P\{|X_k - Y_k| \geq 2(\phi_k + \rho_k)\} \leq P\{|X_k - X_k^*| \geq \tfrac{1}{2}\rho_k\} + P\{|X_k^* - Z_k| \geq 2\phi_k\}$$

$$+ P\{|Z_k - Y_k| \geq \tfrac{3}{2}\rho_k\} \leq 2(\phi_k + \sigma_k). \quad \square$$

3. Proof of Theorem 1.

In section 5 of Kuelbs and Philipp (1977) it was shown that (a) implies (b). Since by Theorem 4.1 of Kuelbs (1977) all three conditions in (b) are equivalent it is enough to prove that (ii) implies (a). This will be carried out in the next four subsections. Following Kuelbs and Philipp (1977) in section 3.1 we approximate the random variables by finite-dimensional ones. In section 3.2 we will define the so-called blocks which in the present context are partial sums of these finite-dimensional random variables. We then estimate the Prohorov distance of the distribution of the properly normalized blocks and the appropriate Gaussian distribution. Here we shall use an argument of Hartman Wintner (1941) and a theorem of Yurinskii (1975). In section 3.3 we prove Proposition 2, a result ostensibly weaker than Theorem 1. In section 3.4 we finally derive Theorem 1 from Proposition 2 using an argument of Major (1976a).

3.1. Approximation by finite-dimensional random variables. Since x_1 is pregaussian there exists a mean zero Gaussian measure μ with covariance structure $T(f,g)$. Denote by Π_N the maps obtained from μ as defined in Lemma 2.1 of Kuelbs (1976). Then by relation (2.4) of Kuelbs (1976)

$$\dim \Pi_N B = \min(N, \dim H_\mu) = d \qquad \text{(say)} \tag{3.1.1}$$

where H_μ is also defined in Lemma 2.1 of Kuelbs (1976). Herc we interpret $\min(N,\infty) = N$. Thus $\Pi_N B$ is a Euclidean space $\mathbb{R}^d$ with the metric induced by the norm in H_μ. We shall denote this metric, as usual, by $|\cdot|$.

Lemma 3.1. Given $\alpha > 0$ there is an N such that

$$\limsup_{n\to\infty} (n \log\log n)^{-\frac{1}{2}} \|\sum_{\nu\le n} (x_\nu - \Pi_N x_\nu)\| < \alpha \quad \text{a.s.}$$

Proof. The proof of Lemma 3.1 is basically the same as the proof of Lemma 4.3 of Kuelbs and Philipp (1977). We assumed that condition (ii) holds. Then by Theorem 4.1 of Kuelbs (1977) condition (iii) holds. We now apply in the proof of Lemma 4.3 of Kuelbs and Philipp (1977) condition (iii) instead of Lemma 4.2 and obtain the lemma. □

The next lemma is Lemma 4.4 of Kuelbs and Philipp (1977).

Lemma 3.2. Let $\{X(t),\ t \geq 0\}$ be a mean zero Brownian motion with covariance structure $T(f,g)$. Then given $\alpha > 0$ there is an N such that

$$\limsup_{t\to\infty}(t \log\log t)^{-\frac{1}{2}}\| X(t) - \Pi_N X(t)\| < \alpha \qquad \text{a.s.}$$

3.2. Definition and approximation of the blocks. We fix $0 < \alpha \leq 1$ and choose N in accordance with Lemmas 3.1 and 3.2. For shorter notation we write

$$\xi_\nu = \Pi_N x_\nu \qquad \nu \geq 1. \tag{3.2.1}$$

Then $\{\xi_\nu,\ \nu \geq 1\}$ is a sequence of independent identically distributed random variables with values in $\mathbb{R}^d$. Moreover, the ξ_ν's are centered at expectations and have finite second moments. We put

$$t_k = [(1+\alpha)^k],\ n_k = t_{k+1} - t_k,\ H_k = [t_k, t_{k+1}) \tag{3.2.2}$$

and

$$X_k = n_k^{-\frac{1}{2}} \sum_{\nu\in H_k} \xi_\nu. \tag{3.2.3}$$

The sums $\sum_{\nu\in H_k}\xi_\nu$ are called the blocks. Let F_k be the distribution of X_k and let G be the distribution of $\Pi_N g$ where g is a random variable with distribution μ. Then G is a Gaussian distribution on $\mathbb{R}^d$ with mean zero and the $d \times d$ identity matrix as its covariance matrix. We also put

$$\rho_k = \alpha^2(\log\log t_k)^{\frac{1}{2}}. \tag{3.2.4}$$

The following proposition in conjunction with Theorem 3 plays a crucial role in the proof of Theorem 1.

Proposition 1. There is a sequence $\{\sigma_k,\ k \geq 1\}$ of nonnegative numbers with

$$\sum_{k\geq 1}\sigma_k < \infty$$

such that

$$F_k(A) \leq G(A^{\rho_k}) + \sigma_k$$

for all Borel sets $A \subset \mathbb{R}^d$.

The proof of Proposition 1 depends on the next three lemmas. We first truncate the random variables ξ_ν using an approach of Hartman and Wintner (1941). Let $\varepsilon(\nu)$ and $\lambda(\nu)$ be defined as on p. 172 of this paper so that $\varepsilon(\nu) \downarrow 0$ and

(3.2.5) $$\lambda(\nu) = (\nu/\log\log \nu)^{\frac{1}{2}}\varepsilon(\nu) \uparrow \infty.$$

We also define

(3.2.6) $$\mu_\nu = E\{|\xi_\nu| I(|\xi_\nu| > \lambda(\nu))\}$$

and recall that by relation (22) of Hartman and Wintner (1941) and its proof

(3.2.7) $$\sum_{\nu \geq 1} \frac{\mu_\nu}{(\nu \log\log \nu)^{\frac{1}{2}}} < \infty.$$

For $\nu \in H_k$ we put

(3.2.8) $$\xi_\nu^* = \xi_\nu I(|\xi_\nu| \leq \lambda(t_{k+1})) - \beta_\nu$$

where we set

(3.2.9) $$\beta_\nu = E\{\xi_\nu I(|\xi_\nu| \leq \lambda(t_{k+1}))\}.$$

We observe that $\{\xi_\nu^*, \nu \in H_k\}$ is a sequence of independent identically distributed random variables with values in $\mathbb{R}^d$ and centered at expectations. We also put

(3.2.10) $$X_k^* = n_k^{-\frac{1}{2}} \sum_{\nu \in H_k} \xi_\nu^*$$

and denote by Γ_k^* the covariance matrix of ξ_ν^*, $\nu \in H_k$.

Lemma 3.3. We have

(3.2.11) $$\sum_{k \geq 1} P\{|X_k^* - X_k| \geq \tfrac{1}{4}\alpha^2 (\log\log t_k)^{\frac{1}{2}}\} < \infty.$$

Moreover, Γ_k^* is the covariance matrix of X_k^* and as $k \to \infty$

(3.2.12) $$\Gamma_k^* \to I.$$

Proof. That Γ_k^* is also the covariance matrix of X_k^* follows at once from the remark after relation (3.2.9). Next we observe that by (3.2.2), (3.2.5), (3.2.6) and (3.2.9)

(3.2.13) $$|\beta_\nu| \le \mu_\nu \downarrow 0$$

since ξ_1 is centered at expectation and has finite second moment. Similarly by (3.2.8) and (3.2.13)

$$E|\xi_\nu^* - \xi_\nu|^2 \le 2E\{|\xi_\nu|^2 I(|\xi_\nu| > \lambda(t_{k+1}))\} + 2|\beta_\nu| \downarrow 0.$$

This implies (3.2.12).

To prove (3.2.11) we note that by (3.2.8) and (3.2.9)

(3.2.14) $$E|\xi_\nu^* - \xi_\nu| \le 2\mu_\nu.$$

Hence by (3.2.2), (3.2.3) and (3.2.10) a typical term of the series in (3.2.11) does not exceed

$$4\alpha^{-2}(n_k \log\log t_k)^{-\frac{1}{2}} \sum_{\nu\in H_k} 2\mu_\nu \ll \sum_{\nu\in H_k} \frac{\mu_\nu}{(\nu \log\log \nu)^{\frac{1}{2}}}$$

which by (3.2.7) is the k-th term of a convergent series. □

Let F_k^* be the distribution of X_k^* and let G_k^* be the Gaussian distribution with mean zero and covariance matrix Γ_k^*. Let $\pi(F_k^*, G_k^*)$ denote the Prohorov distance of these two distributions.

Lemma 3.4. We have

$$\sum_{k\ge 1} \pi(F_k^*, G_k^*) < \infty.$$

Proof. In view of (3.2.12) all eigenvalues $\lambda_{kp}^*(1 \le p \le d)$ of Γ_k^* satisfy $\lambda_{kp}^* \ge \frac{1}{2}$ for $k \ge k_o$. We apply the main theorem of Yurinskii (1975), p. 9 and obtain

(3.2.15) $$\pi(F_k^*, G_k^*) \ll \sum_{p\le d} \frac{\sum_{\nu\in H_k} E|\langle e_{kp}^*, \xi_\nu^*\rangle|^3}{\left\{E(\langle e_{kp}^*, n_k^{\frac{1}{2}} X_k^*\rangle)^2\right\}^{\frac{3}{2}}}$$

where e_{kp}^* $(1 \le p \le d)$ are the eigenvectors of Γ_k^* and where the constant implied by $\ll$ only depends on d and thus only depends on α. We assume that the eigenvectors are normalized to length one. Then by Cauchy's inequality and stationarity

$$(3.2.16)\qquad \sum_{\nu\in H_k} E|\langle e^*_{kp},\ \xi^*_\nu\rangle|^3 \leq n_k E|\xi_{n_k}|^3.$$

For the estimate of the denominator of the right-hand side of (3.2.15) we observe that for $1 \leq p \leq d$

$$E(\langle e^*_{kp},\ X^*_k\rangle)^2 = e^{*\top}_{kp}\, E(X^{*\top}_k X^*_k)\, e^*_{kp} = e^{*\top}_{kp}\, \Gamma^*_k e^*_{kp} = e^{*\top}_{kp}\, \lambda^*_{kp}\, e^*_{kp} = \lambda^*_{kp} \geq \tfrac{1}{2}$$

for $k \geq k_o$. Here $^\top$ denotes the transpose. Thus by (3.2.15) and (3.2.16)

$$(3.2.17)\qquad \pi(F^*_k,\ G^*_k) \ll n_k^{-\frac{1}{2}}\, E|\xi^*_{n_k}|^3$$

where the constant implied by $\ll$ depends on α only.

The lemma follows now by a routine argument. By (3.2.17), (3.2.8), (3.2.13), (3.2.2), (3.2.1) and stationarity

$$\sum_{k\geq 1}\pi(F^*_k,\ G^*_k) \ll \sum_{k\geq 1} n_k^{-\frac{1}{2}}(E\{|\xi_1|^3 I(|\xi_1| \leq \lambda(t_{k+1}))\} + \mu^3_{n_k})$$

$$\ll 1 + \sum_{k\geq 1} n_k^{-\frac{1}{2}} \sum_{j\leq k} n_j^{\frac{3}{2}}\, P\{n_j^{\frac{1}{2}} \leq |\xi_1| < n_{j+1}^{\frac{1}{2}}\}$$

$$\ll 1 + \sum_{j\geq 1} n_j^{\frac{3}{2}}\, P\{n_j^{\frac{1}{2}} \leq |\xi_1| < n_{j+1}^{\frac{1}{2}}\} \sum_{k\geq j} n_k^{-\frac{1}{2}}$$

$$\ll 1 + \sum_{j\geq 1} n_j P\{n_j^{\frac{1}{2}} \leq |\xi_1| < n_{j+1}^{\frac{1}{2}}\} \ll 1 + E|\xi_1|^2 \ll 1. \square$$

<u>Lemma 3.5</u>. We have for all Borel sets $A \subset \mathbb{R}^d$

$$G(A) \leq G^*_k(A^{\frac{1}{4}\rho_k}) + O(k^{-3/2}).$$

<u>Proof</u>. In view of (3.2.12) the matrix Γ^*_k is nonsingular for $k \geq k_o$. Hence it is enough to show that

$$(3.2.18)\qquad \int_A e^{-\frac{1}{2}u^2}du \leq (\det \Gamma^*_k)^{-\frac{1}{2}}\int_{A^{\frac{1}{4}\rho_k}} \exp(-\tfrac{1}{2}\langle u,\Gamma^{*-1}_k u\rangle)du + O(k^{-3/2})$$

for all Borel sets $A \subset \mathbb{R}^d$. By a simple change of variable (3.2.18) transforms to

$$\int_A e^{-\frac{1}{2}u^2}du \le \int_{\Gamma_k^{*-\frac{1}{2}}(A^{\frac{1}{4}\rho_k})} e^{-\frac{1}{2}u^2}du + O(k^{-3/2}). \tag{3.2.19}$$

Since the tail of a Gaussian random variable tends to zero square exponentially (for the relevant estimate see for instance relation (2.26) of Berkes and Philipp (1979)) it is enough to prove (3.2.19) for all Borel sets $A \subset \{x \in \mathbb{R}^d : |x| \le 2(\log\log t_k)^{\frac{1}{2}}\}$. But to prove (3.2.19) for such Borel sets A it is enough to observe that by (3.2.12)

$$|I - \Gamma_k^{*\frac{1}{2}}| \le \frac{1}{2}\alpha^2 \tag{3.2.20}$$

for all $k \ge k_o$. Here $|\cdot|$ denotes the norm of a matrix. Indeed, by (3.2.20) we obtain for each u in such a bounded Borel set A

$$|u - \Gamma_k^{*\frac{1}{2}}u| \le \frac{1}{8}\alpha^2(2\log\log t_k)^{\frac{1}{2}} = \tfrac{1}{4}\rho_k$$

which implies $\Gamma_k^{*\frac{1}{2}}u \in A^{\frac{1}{4}\rho_k}$. Consequently $A \subset \Gamma_k^{*-\frac{1}{2}}(A^{\frac{1}{4}\rho_k})$ and this yields (3.2.19). □

We now can finish the proof of Proposition 1. By Proposition 1 of Dudley (1968) Lemma 3.5 is equivalent to

$$G_k^*(A) \le G(A^{\frac{1}{4}\rho_k}) + O(k^{-3/2})$$

for all Borel sets $A \subset \mathbb{R}^d$. Hence by Lemmas 3.3, 3.4 and 3.5 we have for each Borel set $A \subset \mathbb{R}^d$

$$\begin{aligned} F_k(A) &= P(X_k \in A) \le P(X_k^* \in A^{\frac{1}{4}\rho_k}) + \tfrac{1}{4}\sigma_k \\ &= F^*(A^{\frac{1}{4}\rho_k}) + \tfrac{1}{4}\sigma_k \le G_k^*(A^{\frac{1}{4}\rho_k+\frac{1}{4}\sigma_k}) + \tfrac{1}{2}\sigma_k \\ &\le G(A^{\rho_k}) + \sigma_k \end{aligned}$$

with $\sum_{k\ge 1}\sigma_k < \infty$.

3.3. Proof of Proposition 2.

As already mentioned above we shall derive Theorem 1 from the following proposition.

Proposition 2. Let $0 < \alpha \leq 10^{-4}$. Without changing its distribution we can redefine the sequence $\{x_\nu, \nu \geq 1\}$ on a new probability space on which there exists Brownian motion $\{X(t), t \geq 0\}$, depending perhaps on α, with covariance structure $T(f,g)$ such that

$$\limsup_{t\to\infty} (t \log\log t)^{-\frac{1}{2}} \left| \sum_{\nu\leq t} x_\nu - X(t) \right| \leq \alpha \qquad \text{a.s.}$$

Let N be so large that Lemmas 3.1 and 3.2 hold. For the proof of Proposition 2 we need the following two lemmas.

Lemma 3.6. We have

$$\limsup_{m\to\infty} (t_m \log\log t_m)^{-\frac{1}{2}} \max_{t_m<n\leq t_{m+1}} \left| \sum_{\nu=t_m+1}^{n} \xi_\nu \right| \leq 2\alpha^{\frac{1}{4}} \qquad \text{a.s.}$$

Proof. We apply Ottaviani's inequality which remains valid in the Banach space setting. For its statement and proof see e.g. Breiman (1968), p. 45. We obtain

$$(3.3.1) \qquad P\{\max_{t_m\leq n\leq t_{m+1}} \left| \sum_{\nu=t_m+1}^{n} \xi_\nu \right| > 2\alpha^{\frac{1}{4}} (t_m \log\log t_m)^{\frac{1}{2}}\}$$

$$\leq (1-c)^{-1} P\{ \left| \sum_{\nu\in H_m} \xi_\nu \right| > \alpha^{\frac{1}{4}} (t_m \log\log t_m)^{\frac{1}{2}}\}$$

where by stationarity, (3.2.2) and Chebyshev's inequality

$$(3.3.2) \quad c = c_m = \max_{t_m<n\leq t_{m+1}} P\{ \left| \sum_{\nu\leq n-t_m} \xi_\nu \right| > \alpha^{\frac{1}{4}} (t_m \log\log t_m)^{\frac{1}{2}}\}$$

$$\leq \max_{t_m<n\leq t_{m+1}} P\{ \left| \sum_{\nu\leq n-t_m} \xi_\nu \right| > \tfrac{1}{2}\alpha^{-\frac{1}{4}} (n_m \log\log n_m)^{\frac{1}{2}}\} = o(1).$$

By Proposition 1, (3.2.2) and (3.2.3) we obtain

$$\sum_{k\geq 1} P\{ \left| \sum_{\nu\in H_k} \xi_\nu \right| > \alpha^{\frac{1}{4}} (t_k \log\log t_k)^{\frac{1}{2}}\}$$

$$\leq \sum_{k\geq 1} F_k\{x \in \mathbb{R}^d : |x| > \tfrac{1}{2}\alpha^{-\frac{1}{4}} (\log\log t_k)^{\frac{1}{2}}\}$$

$$\ll 1 + \sum_{k\geq 1} G\{x \in \mathbb{R}^d : |x| \geq \tfrac{1}{4}\alpha^{-\frac{1}{4}}(\log\log t_k)^{\frac{1}{2}}\} \ll 1$$

since G is a Gaussian distribution and $\alpha \leq 10^{-4}$. Hence by (3.3.1), (3.3.2) and the Borel Cantelli Lemma there is with probability 1 a $m_o(\omega)$ such that for all $m \geq m_o$

$$\max_{t_m<n\leq t_{m+1}} |\sum_{\nu=t_m+1}^{n} \xi_\nu| \leq 2\alpha^{\frac{1}{4}}(t_m \log\log t_m)^{\frac{1}{2}}.$$

We divide by $(t_m \log\log t_m)^{\frac{1}{2}}$ take the limes superior $(m \to \infty)$ and obtain the lemma. □

Lemma 3.7. We have for any Brownian motion $\{X(t),\ t \geq 0\}$ with covariance structure $T(f,g)$

$$\limsup_{m\to\infty} (t_m \log\log t_m)^{-\frac{1}{2}} \max_{t_m\leq t\leq t_{m+1}} |\Pi_N X(t) - \Pi_N X(t_m))| \leq 2\alpha^{\frac{1}{4}} \quad \text{a.s.}$$

Proof. This follows easily from standard facts for Brownian motion. Indeed, since $\{\Pi_N X(t),\ t \geq 0\}$ is standard Brownian motion in $\mathbb{R}^d$ we have by (3.2.2)

$$P\{\max_{t_m\leq t\leq t_{m+1}} |\Pi_N(X(t) - X(t_m))| \geq \alpha^{\frac{1}{4}}(t_m \log\log t_m)^{\frac{1}{2}}\}$$

$$\ll P\{|\Pi_N(X(t_{m+1}) - X(t_m))| \geq \tfrac{1}{2}\alpha^{\frac{1}{4}}(t_m \log\log t_m)^{\frac{1}{2}}\}$$

$$\ll P\{|\Pi_N X(1)| \geq \tfrac{1}{4}\alpha^{-\frac{1}{4}}(\log\log t_m)^{\frac{1}{2}}\}$$

$$\ll G\{|x| \geq \tfrac{1}{2}\alpha^{\frac{1}{4}}(\log\log t_m)^{\frac{1}{2}}\} \ll m^{-2}.$$

The lemma follows now from the Borel-Cantelli lemma as the previous one. □

We now can finish the proof of Proposition 2. By Proposition 1 and by Theorem 3 with $\phi_k = 0$ we can redefine the sequences $\{X_k,\ k \geq 1\}$ and $\{x_\nu,\ \nu \geq 1\}$ without changing their joint distribution on a new probability space on which there exists a sequence $\{Y_k,\ k \geq 1\}$ of independent Gaussian random variables with common distribution G such that

$$\sum_{k\geq 1} P\{|X_k - Y_k| \geq 2\alpha^2(\log\log t_k)^{\frac{1}{2}}\} < \infty. \tag{3.3.3}$$

Hence by the Borel Cantelli lemma and by (3.2.3) there is with probability 1 a $k_1(\omega)$ such that for all $k \geq k_1(\omega)$

$$(3.3.4) \qquad |\sum_{\nu\in H_k}\xi_\nu - n_k^{\frac{1}{2}}Y_k| \leq 2\alpha^2(n_k \log\log t_k)^{\frac{1}{2}}.$$

Let $\{X(t),\ t \geq 0\}$ be any Brownian motion in B with covariance structure $T(f,g)$. Then by (3.2.2) $n_k^{-\frac{1}{2}}(X(t_{k+1}) - X(t_k))$ has distribution μ. Thus $n_k^{-\frac{1}{2}}(\Pi_N X(t_{k+1}) - \Pi_N X(t_k))$ has distribution G since Π_N is linear. Consequently the sequences $\{n_k^{-\frac{1}{2}}(\Pi_N X(t_{k+1}) - \Pi_N X(t_k)),\ k \geq 1\}$ and $\{Y_k,\ k \geq 1\}$ have the same distribution. Hence by Kolmogorov's existence theorem which remains valid in the Banach space setting, we can redefine the process $\{\sum_{\nu\leq t} x_\nu,\ t \geq 0\}$ and the sequence $\{Y_k,\ k \geq 1\}$ on a richer probability space without changing their joint distribution such that on this probability space there exists Brownian motion $\{X(t),\ t \geq 0\}$ with covariance structure $T(f,g)$ and satisfying

$$(3.3.5) \qquad n_k^{-\frac{1}{2}}(\Pi_N X(t_{k+1}) - \Pi_N X(t_k)) = Y_k.$$

We now show that $\{X(t),\ t \geq 0\}$ has the desired properties. Let $t > 0$ be given and define m by $t_m \leq t < t_{m+1}$. Then by (3.2.1), (3.2.2), (3.3.4), (3.3.5) we obtain

$$(3.3.6) \qquad \limsup_{m\to\infty}(t_m \log\log t_m)^{-\frac{1}{2}}|\sum_{\nu\leq t_m}\xi_\nu - \Pi_N X(t_m)| < \alpha \qquad \text{a.s.}$$

By relation (2.3) of Kuelbs (1976) we have $\|x\| < c\|x\|_{H_\mu}$ for $x \in H_\mu$ and by relation (2.4) of Kuelbs (1976) we have $\|y\|_{H_\mu} = |y|$ for $y \in \Pi_N B = \Pi_N H_\mu = \mathbb{R}^d$. Here $c \geq 1$ is a constant. Consequently,

$$\begin{aligned}\|\sum_{\nu\leq t} x_\nu - X(t)\| &\leq \|\sum_{\nu\leq t}(x_\nu - \xi_\nu)\| + c\max_{t_m<n\leq t_{m+1}}|\sum_{\nu=t_m+1}^{n}\xi_\nu| \\ &+ c\,|\sum_{\nu\leq t_m}\xi_\nu - \Pi_N X(t_m)| \\ &+ c\max_{t_m\leq t\leq t_{m+1}}|\Pi_N X(t_m) - \Pi_N X(t)| + \|\Pi_N X(t) - X(t)\|.\end{aligned}$$

We divide by $(t \log\log t)^{\frac{1}{2}}$, take the limes superior as $t \to \infty$, apply (3.3.6) and Lemmas 3.6, 3.7, 3.1 and 3.2 and obtain the bound $14\alpha^{\frac{1}{4}}c$.

3.4. Conclusion of the proof of Theorem 1.

We need one more lemma.

Lemma 3.8. Let $\{m_k, k \geq 1\}$ be any increasing sequence of integers with $m_k \geq 2^k$. Put $r_k = 2^{m_k}$. Then for any $\delta > 0$

$$\sum_{k\geq 1} P\{\| \sum_{\nu\leq r_k} x_\nu \| \geq \delta (r_k \log\log r_k)^{\frac{1}{2}}\} < \infty.$$

Proof. We write for $k \geq 1$ and $\nu \leq 2^k$

$$x_{\nu k}^* = x_\nu I(\| x_\nu \| < \lambda(2^k)) - E\{x_\nu I(\| x_\nu \| < \lambda(2^k))\}.$$

Then by stationarity and (3.2.6)

$$E\| x_\nu - x_{\nu k}^* \| \leq 2\mu_{2^k}. \tag{3.4.1}$$

Hence by (3.2.7) and (3.2.13)

$$\begin{aligned}
&\sum_{k\geq 1} P\{\| \sum_{\nu\leq r_k} (x_\nu - x_{\nu m_k}^*) \| \geq \delta (r_k \log\log r_k)^{\frac{1}{2}}\} \\
&\leq \sum_{k\geq 1} P\{\| \sum_{\nu\leq 2^k} (x_\nu - x_{\nu k}^*) \| \geq \delta (2^k \log\log 2^k)^{\frac{1}{2}}\} \\
&\ll \sum_{k\geq 1} 2^k \frac{\mu_{2^k}}{(2^k \log\log 2^k)^{\frac{1}{2}}} \ll \sum_{n\geq 1} \frac{\mu_n}{(n \log\log n)^{\frac{1}{2}}} < \infty.
\end{aligned} \tag{3.4.2}$$

We now apply Lemma 2.1 of Kuelbs (1977) to the sequence $\{x_{\nu m_k}^*, \nu \leq r_k\}$ for fixed $k \geq 1$. We put $b_{r_k} = r_k^{\frac{1}{2}}$, $c = \varepsilon(r_k)(\log\log r_k)^{-\frac{1}{2}}$ and $\varepsilon = \frac{1}{2}\delta(\log\log r_k)^{\frac{1}{2}}$. We observe that by (3.4.1), (3.2.7) and condition (ii) of statement (b)

$$\begin{aligned}
E\| \sum_{\nu\leq r_k} x_{\nu m_k}^* \| &\leq \sum_{\nu\leq r_k} E\| x_\nu - x_{\nu m_k}^* \| + E\| \sum_{\nu\leq r_k} x_\nu \| \\
&\leq \sum_{\nu\leq r_k} \mu_\nu + o((r_k \log\log r_k)^{\frac{1}{2}}) = o((r_k \log\log r_k)^{\frac{1}{2}})
\end{aligned}$$

Hence

$$P\{\| \sum_{\nu\leq r_k} x_{\nu m_k}^* \| \geq \delta (r_k \log\log r_k)^{\frac{1}{2}}\} \leq \exp(-\delta^2 k/100).$$

The lemma follows now from (3.4.2) and the last estimate. □

We finally can finish the proof of Theorem 1 using an argument similar to Major (1976a), pp. 223-224. By Proposition 2 there is for each $k \geq 1$ a sequence $\{x_\nu^{(k)}, \nu \geq 1\}$ of random variables with the same distribution as $\{x_\nu, \nu \geq 1\}$ and a Brownian motion $\{X_k(t), t \geq 0\}$ with covariance structure $T(f,g)$ such that with probability 1

$$(3.4.3) \qquad \limsup_{t\to\infty} (t \log\log t)^{-\frac{1}{2}} \| \textstyle\sum_{\nu\leq t} x_\nu^{(k)} - X_k(t) \| \leq 2^{-k-1}.$$

Without loss of generality we can assume that the sequences $\{x_\nu^{(k)}, \nu \geq 1\}$ and the Brownian motions $\{X_k(t), t \geq 0\}$ are all defined on the same probability space and are independent. From (3.4.3) and Egorov's theorem we conclude that there exists a sequence $\{r_k, k \geq 1\}$ with $r_k = 2^{m_k}$ such that

$$(3.4.4) \qquad P\{\sup_{t\geq r_k} (t \log\log t)^{-\frac{1}{2}} \| \textstyle\sum_{\nu\leq t} x_\nu^{(k)} - X_k(t) \| \geq 2^{-k}\} \leq 2^{-k}.$$

We now construct inductively the desired Brownian motion $\{X(t), t \geq 0\}$ by setting $r_{-1} = 0$ and

$$(3.4.5) \qquad X(t) = X(r_k) + X_k(t) - X_k(r_k) \quad \text{if} \quad r_k < t \leq r_{k+1}, \quad k \geq -1$$

and a sequence $\{y_\nu, \nu \geq 1\}$ of random variables by setting

$$(3.4.6) \qquad y_\nu = x_\nu^{(k)} \quad \text{if} \quad r_k < \nu \leq r_{k+1}, \quad k \geq -1.$$

Then $\{y_\nu, \nu \geq 1\}$ as well as $\{x_\nu^{(k)}, \nu \geq 1\}$ have the same distribution as $\{x_\nu, \nu \geq 1\}$. Consequently by Lemma 3.8 and the Borel Cantelli Lemma we have

$$(3.4.7) \qquad \| \textstyle\sum_{\nu\leq r_k} y_\nu \| = o((r_k \log\log r_k)^{\frac{1}{2}}) \qquad \text{a.s.}$$

and

$$(3.4.8) \qquad \| \textstyle\sum_{\nu\leq r_k} x_\nu^{(k)} \| = o((r_k \log\log r_k)^{\frac{1}{2}}) \qquad \text{a.s.}$$

Applying Lemma 3.8 to the increments of $\{X(t), t \geq 0\}$ and $\{X_k(t), t \geq 0\}$ we obtain

$$(3.4.9) \qquad \| X(r_k) \| = o((r_k \log\log r_k)^{\frac{1}{2}}) \qquad \text{a.s.}$$

and

(3.4.10) $$\| X_k(r_k) \| = o((r_k \log\log r_k)^{\frac{1}{2}}) \qquad \text{a.s.}$$

Moreover, by (3.4.4)

(3.4.11) $$\sup_{t \geq r_k} (t \log\log t)^{-\frac{1}{2}} \| \sum_{\nu \leq t} x_\nu^{(k)} - X_k(t) \| = o(1) \qquad \text{a.s.}$$

Theorem 1 follows now from (3.4.5) - (3.4.11).

4. Proof of Corollary 4.

In the absence of the assumption that x_1 be pregaussian the existence of a Brownian motion $\{X(t),\ t \geq 0\}$ with covariance structure $T(f,g)$ is no longer guaranteed. But Corollary 4 follows at once from the proof of Proposition 2. We omit Lemma 3.2, interpret the process $\{\Pi_N X(t),\ t \geq 0\}$ as $\{X_\alpha(t),\ t \geq 0\}$ in Lemma 3.7 and in the argument following it and obtain (1.3). Condition (1.2) can also be proved rather easily.

Although Corollary 4 contains Theorem 4.1 of Kuelbs (1977) as a special case this does not provide a new proof of it since in the proof of Lemma 3.1 we used the fact that (ii) implies (iii), thus using Kuelbs' result in a crucial way.

In the remainder of this section we shall give a proof of Corollary 4 which in essence does not depend on Theorem 4.1 of Kuelbs (1977) although it is based on several ingredients used by Kuelbs in his proof of Theorem 4.1. That (iii) implies (ii) was shown by Kuelbs (1977), pp. 795-796. The implication (ii) $\rightarrow$ (i) is trivial. Here we show first that (i) implies (iv) and later that (iv) implies (iii).

To show that (i) implies (iv) we observe that for the proof of Corollary 1 Lemmas 3.1 and 3.2 are not needed. Hence it is enough to approximate the random variables x_ν by finite-dimensional ones arbitrarily closely. To be precise let $\alpha > 0$ and let τ_α be the mapping with finite-dimensional range as defined by Kuelbs (1977), p. 790. We put for $\nu \geq 1$

$$x'_\nu = \tau_\alpha x_\nu .$$

Then

(4.1) $$Ex'_\nu = 0 \quad \text{and} \quad E\|x_\nu - x'_\nu\|^2 \le \alpha.$$

With this notation we have the following bounded law of the iterated logarithm.

Proposition 3. Suppose that condition (i) holds. Then

$$\limsup_{n\to\infty}(2n \log\log n)^{-\frac{1}{2}}\|\sum_{\nu\le n}(x_\nu - x'_\nu)\| \le 8\alpha^{\frac{1}{2}} \quad \text{a.s.}$$

For the proof we write

$$z_\nu = x_\nu - x'_\nu$$

and truncate the random variables z_ν as in section 3.2. We define

$$z^*_\nu = z_\nu I(\|z_\nu\| \le \lambda(\nu)) - E\{z_\nu I(\|z_\nu\| \le \lambda(\nu))\}.$$

Then by the argument of Hartman and Wintner (1941), pp. 172-176 which remains valid in the Banach space setting we obtain

(4.2) $$\limsup_{n\to\infty}(n \log\log n)^{-\frac{1}{2}}\|\sum_{\nu\le n}(z_\nu - z^*_\nu)\| = 0 \quad \text{a.s.}$$

We now apply Theorem 3.1 of Kuelbs (1977) to the sequence $\{z^*_\nu, \nu \ge 1\}$ with $\sigma_n^2 = \alpha n$. From the second of the remarks on p. 789 of Kuelbs' paper we obtain

(4.3) $$\limsup_{n\to\infty}(n \log\log n)^{-\frac{1}{2}}\|\sum_{\nu\le n} z^*_\nu\| \le 8\alpha^{\frac{1}{2}} \quad \text{a.s.}$$

The proposition follows now from (4.2) and (4.3). □

We apply Corollary 1 to the sequence $\{x'_\nu, \nu \ge 1\}$ and obtain without loss of generality a Brownian motion $\{X_\alpha(t), t \ge 0\}$ satisfying

(4.4) $$\sum_{\nu\le t} x'_\nu - X_\alpha(t) = o((t \log\log t)^{\frac{1}{2}}) \quad \text{a.s.}$$

Since in finite-dimensional spaces all norms are equivalent relation (1.3) follows from (4.4) and Proposition 3. Moreover, (1.2) follows easily by (4.1) and an argument similar to the one used to prove (3.2.15).

To complete the cycle of implications we show now that (iv) implies (iii). This is essentially known. Indeed, the sequence

$\{(2n \log\log n)^{-\frac{1}{2}} X_\alpha(n),\ n \geq 1\}$ satisfies the compact law of the iterated logarithm with cluster set K_α where K_α is the unit ball of H_α, the Hilbert space determined by T_α. Hence by (1.3) the sequence $\{(2n \log\log n)^{-\frac{1}{2}} \sum_{\nu \leq n} x_\nu,\ \nu \geq 1\}$ clusters in a set K^* satisfying $K^* \subset K_\alpha^\varepsilon$ and $K_\alpha \subset K^{*\varepsilon}$ for any $\varepsilon > 0$. But by (1.2) and by relation (2.2) of Kuelbs (1976) the unit ball K of the Hilbert space H determined by T also satisfies $K \subset K_\alpha^\alpha$ and $K_\alpha \subset K^\alpha$ for any $\alpha > 0$. Hence $K^* = K$.

5. Proof of Corollary 3.

By hypothesis

$$n^{-\frac{1}{2}} \sum_{\nu \leq n} x_\nu \longrightarrow \lambda \qquad \text{in distr.}$$

Hence we obtain for any $f \in B^*$

$$n^{-\frac{1}{2}} \sum_{\nu \leq n} f(x_\nu) \longrightarrow \lambda \circ f^{-1} \qquad \text{in distr.}$$

Thus $E\{f^2(x_1)\} < \infty$. Consequently, x_1 has a covariance structure $T(f,g)$, say, and, as is easy to see, $T(f,g)$ is also the covariance structure of λ. Let K denote the unit ball of H_λ, as defined in Lemma 2.1 of Kuelbs (1976) and let K^* denote the cluster set of $\{(2n \log\log n)^{-\frac{1}{2}} \sum_{\nu \leq n} x_\nu,\ n \geq 1\}$ whose existence is guaranteed by the second hypothesis of Corollary 3. We apply Theorem 3.1 of Kuelbs (1976) with $Y_n = \sum_{\nu \leq n} x_\nu$ and $\phi_n = (2n \log\log n)^{\frac{1}{2}}$ and obtain $K = K^*$. Indeed, since $\{x_\nu,\ \nu \geq 1\}$ satisfies the compact law of the iterated logarithm we obtain for each $f \in B^*$

$$\limsup_{n\to\infty} (2n \log\log n)^{-\frac{1}{2}} \sum_{\nu \leq n} f(x_\nu) = E^{\frac{1}{2}}\{f^2(x_1)\} = T^{\frac{1}{2}}(f,f)$$

$$= \sup_{x \in K} f(x)$$

by relation (2.5) of Kuelbs (1976).

Let Π_N be the maps obtained from λ as defined in Lemma 2.1 of Kuelbs (1976). Since $K = K^*$ Lemma 3.1 and its proof remain valid as they stand. Since $\{\Pi_N x_\nu, \nu \geq 1\}$ is a sequence of independent identically distributed random variables with values in $\mathbb{R}^d$ satisfying a central limit theorem the random variables $\Pi_N x_\nu$ have finite second moments. As we go along we reinterpret in the proof of Theorem 1 $\{X(t), t \geq 0\}$ as a Brownian motion determined by λ and obtain Corollary 3.

6. Proof of Theorem 2.

The proof is based on the same ideas but is much simplier than the proof of Theorem 1. We also need the following proposition, essentially due to R. N. Bhattacharya (1978). Before stating this result we make some simplifications. By passing to a suitable subspace of $\mathbb{R}^d$, if necessary, we observe that there is no loss of generality if we assume that first Γ is nonsingular and that second Γ is the $d \times d$ identity matrix. This always can be achieved by subjecting the random variables to the linear transformation $x \to \Gamma^{-\frac{1}{2}}x$. Of course, another way of doing this is to apply Lemma 2.1 of Kuelbs (1976).

Proposition 4. Let G_n be the distribution of $n^{-\frac{1}{2}} \sum_{\nu \leq n} x_\nu$ and let G be the multivariate normal distribution with the identity matrix as covariance matrix. Then

$$\pi(G_n, G) \ll n^{-\frac{1}{2}\delta} \log n.$$

For $\delta = 1$ this is just Theorem 19.7 on pp. 174-175 of R. N. Bhattacharya and Rango Rao (1975). For $0 < \delta < 1$ Bhattacharya (1978) obtained this result form the case $\delta = 1$ using a truncation argument. Of course, for $\delta = 1$, Yurinskii's (1975) theorem is stronger. It is actually best possible. It is very likely that in the case $0 < \delta < 1$ the factor $\log n$ also can be omitted. This would result in a slight reduction of the error term in (1.2), namely to $\ll t^{1/(2+\delta)} \log^{\frac{1}{2}} t$. One might wonder whether Major's (1976b) bound $o(t^{1/(2+\delta)})$ could be obtained by means of the present method.

Let $\varepsilon > 0$. We define

$$t_k = \left[k^{(2+\delta)/\delta}(\log k)^{(4+\varepsilon)/\delta}\right], \quad n_k = t_{k+1} - t_k, \quad H_k = [t_k, t_{k+1})$$

and

$$X_k = n_k^{-\frac{1}{2}} \sum_{\nu \in H_k} x_\nu .$$

Let F_k be the distribution of X_k. Then by Proposition 4

$$\pi(F_k, G) \ll k^{-1}(\log k)^{-1-\frac{1}{2}\varepsilon}$$

an estimate corresponding to Proposition 1. Hence by Theorem 3 we can redefine the sequence $\{X_k, k \geq 1\}$ on a new probability space on which there exists a sequence $\{Y_k, k \geq 1\}$ of independent Gaussian random variables with common distribution G such that

$$P\{|X_k - Y_k| \geq k^{-1}(\log k)^{-1-\frac{1}{4}\varepsilon}\} \ll k^{-1}(\log k)^{-1-\frac{1}{2}\varepsilon}.$$

The arguments yielding Lemmas 3.6 and 3.7 give

$$\max_{t_m < n \leq t_{m+1}} \left|\sum_{\nu=t_m+1}^{n} x_\nu\right| \ll t_m^{1/(2+\delta)} \log t_m \qquad \text{a.s.}$$

and

$$\max_{t_m \leq t \leq t_{m+1}} |X(t) - X(t_m)| \ll t_m^{1/(2+\delta)} \log t_m \qquad \text{a.s.}$$

The proof of Theorem 2 can finally be finished as in section 3.3.

Acknowledgment. I am grateful to Rabi Bhattacharya for showing me his proof of Proposition 4.

References

Berkes, István and Walter Philipp (1979), Approximation theorems for independent and weakly dependent random vectors, Annals of Probability, 7,

Bhattacharya, R. N. (1978), private communication.

Bhattacharya, R. N. and R. Ranga Rao (1975), Normal approximation and asymptotic expansion, Wiley, New York.

Breiman, Leo (1968), Probability, Addison Wesley, Reading, Mass.

Dudley, R. M. (1968), Distances of probability measures and random variables, Annals Math. Stat. 39, 1563-1572.

Hartman, Philip and Aurel Wintner (1941), On the law of the iterated logarithm, Amer. J. Math. 63, 169-176.

Heinkel, B. (1979), Sur la loi du logarithme itéré pour des v.a. à valeurs dans un espace de Banach, this volume.

Kuelbs, J. (1976), A strong convergence theorem for Banach space valued random variables, Annals of Probability 5, 744-771.

Kuelbs, J. (1977), Kolmogorov's law of the iterated logarithm for Banach space valued random variables, Illinois J. Math. 21, 784-800.

Kuelbs, J. and Walter Philipp (1977), Almost sure invariance principles for partial sums of mixing B-valued random variables, preprint.

Major, Peter (1976a), Approximation of partial sums of i.i.d.r.v.s. when the summands have only two moments, Z. Wahrscheinlichkeitstheorie verw. Geb. 35, 221-229.

Major, Peter (1976b), The approximation of partial sums of independent r.v.s., Z. Wahrscheinlichkeitstheorie verw. Geb. 35, 213-220.

Philipp, Walter (1978), Weak invariance principles for sums of B-valued random variables, preprint.

Pisier, G. (1975), Le théorème de la limit centrale et la loi du logarithme itéré dans les espace de Banach, Séminaire Maurey-Schwartz 1975-1976, Exposé III, Ecole Polytechnique, Paris.

Strassen, V. (1964), An almost sure invariance principle for the law of the iterated logarithm, Z. Wahrscheinlichkeitstheorie verw. Geb. 3, 211-226.

Strassen, V. (1965), The existence of probability measures with given marginals, Annals Math. Stat. 36, 423-439.

Yurinskii, V. V. (1975), A smoothing inequality for estimates of the Lévy-Prokhorov distance, Theory. Prob. Appl. 20, 1-10.

HILBERTIAN SUPPORT OF A PROBABILITY MEASURE ON A BANACH SPACE

Hiroshi SATO

Department of Mathematics
Kyushu University-33
Hakozaki Fukuoka
812-*Japan*

SUMMARY

In this paper, we will prove the following : Let E be a real separable Banach space. Then every probability measure on E has a Hilbertian support if and only if E is isomorphic to a Hilbert space. In the case of ℓ_p $(1 \leqslant p < 2)$ we will give an explicit construction of probability measures without Hilbertian support.

§1. INTRODUCTION AND DEFINITIONS

Let E be a real separable Banach space, $B(E)$ be the Borel field, μ be a probability measure on $E = (E, B(E))$, E' be the topological dual space of E, and denote the canonical bilinear form on $E \times E'$ by $\langle x,\xi\rangle$. Then μ is said to have a *Hilbertian support* if there exists a continuous injection ψ of a Hilbert space H into E such that $\mu(\psi(H)) = 1$.

R.M. Dudley ([4] p318) and H. Sato ([9] Example 2) showed the existence of a Gaussian measure on a Banach space without Hilbertian support. B. Maurey ([6] Theorem 2) showed that if E is a Banach space of cotype 2, every centered Gaussian measure on E has a Hilbertian support. O.G. Smolyanov and A.V. Uglanov [11] showed that the Wiener measure on C[0,1] has no Hilbertian support. B. Diallo [3] showed that there exists a probability measure without Hilbertian support on ℓ_1.

If there exists a Banach space E where every probability measure has a Hilbertian support, then the analysis of probability measures on E shall be reduced to that on a Hilbert space. But such a Banach space is isomorphic to a Hilbert space and the aim of this paper is to prove the following theorem by utilizing the Mouchtari's lemma.

THEOREM *Let* E *be a real separable Banach space. Then the following two conditions are equivalent.*

(1°) *Every probability measure on* E *has a Hilbertian support*

(2°) E *is isomorphic to a Hilbert space.*

As an application of the theorem, we will construct explicitly a probability measure without Hilbertian support on ℓ_p, $1 \leq p < 2$.

To begin with, we remark the following definitions which will be used later.

DEFINITION 1. A *non-negative definite functional* $\chi(\xi)$ on E' is a non-negative definite functional on E' with $\chi(0) = 1$.

DEFINITION 2. An S-*topology* on E' is a vector topology τ on E' such that the τ-continuity of a non-negative definite functional $\chi(\xi)$ is equivalent to the existence of a probability measure μ on E such that

$$\chi(\xi) = \int_E e^{i<x,\xi>} d\mu(x), \qquad \xi \in E'.$$

We denote by $\tilde{\mu}(\xi)$ the right hand side, and call it the characteristic functional of μ.

DEFINITION 3. A *standard Bernoulli sequence* $\{\varepsilon_n\}$ is an independent random sequence with the same distribution

$$P(\varepsilon_n = 1) = P(\varepsilon_n = -1) = \frac{1}{2}$$

DEFINITION 4. A Banach space E is of *cotype* 2 if for every sequence $\{x_n\}$ in E such that $\sum_n \varepsilon_n x_n$ converges almost surely, $\sum_n \|x_n\|^2$ converges.

DEFINITION 5. An S-*operator* on a Hilbert space H is a non-negative symmetric nuclear operator.

§2. PROOF OF THEOREM

In the proof of the theorem, we make use of the following lemmas.

LEMMA 1. *Let* E *be a real separable Banach space where every probability measure has a Hilbertian support. Then* E *is of cotype* 2.

(PROOF) Let $\{x_n\}$ be a sequence in E for which $\sum_n \varepsilon_n x_n$ converges almost surely and μ be the limit distribution on E, where $\{\varepsilon_n\}$ is a standard Bernoulli sequence. Then, by assumption, there exists a continuous injection ψ of a Hilbert space H into E for which $\mu(\psi(H)) = 1$. Since E is separable, H is necessarily separable (P. Baxendale [1], Lemma 1-3) so that $\psi(H)$ is a Borel subset of E (K.R. Parthasarathy [8], Chap. 1,

Corollary 3-3). Since ψ is injective and continuous, we may identify H with $\psi(H)$, and there exists a positive constant K such that

$$\|x\| \leq K\|x\|_H, \qquad x \in H,$$

where $\|x\|$, $\|x\|_H$ is the norm of E, H, respectively.

On the other hand,

$$P(\sum_n \varepsilon_n x_n \in H) = \mu(H) = 1$$

implies that the sequence $\{x_n\}$ is included in H. In fact, assume that x_n $(\neq 0)$ is not in H, then by the independence of E-valued random variables $\varepsilon_n x_n$ and $Y = \sum_{m \neq n} \varepsilon_m x_m$ we have

$$1 = P(\sum_m \varepsilon_m x_m \in H)$$

$$= \frac{1}{2}P(Y \in H+x_n) + \frac{1}{2}P(Y \in H-x_n)$$

so that

$$P(Y \in H+x_n) = P(Y \in H-x_n) = 1.$$

However, since H is a linear subspace of E, $H+x_n$ and $H-x_n$ are disjoint and this is a contradiction.

Consequently, since the Hilbert space H is of cotype 2,

$$P(\sum_n \varepsilon_n x_n \text{ converges in } H) = 1$$

implies

$$\|x_n\|^2 \leq K^2 \|x_n\|_H^2 < +\infty.$$

This proves the lemma.

LEMMA 2. (D. Mouchtari [7], Cor.2) *Let* E *be a real separable Banach space and assume that there exists an* S-*topology on* E' *induced by a family of non-negative definite symmetric bilinear forms on* E' × E'. *Then* E *is isomorphic to a Hilbert space.*

(**PROOF OF THEOREM**) (2°) → (1°) is trivial.

In order to prove (1°) → (2°), we have only to show that (1°) implies the assumption of Lemma 2.

Let E be a real separable Banach space and μ be a probability measure on E. Then by assumption there exists a continuous injection ψ of a Hilbert space H into E. Since H is separable, the image $\psi(A)$ of every Borel subset A of H is a Borel subset of E, and we induce a probability measure μ_H on H by

$$\mu_H(A) = \mu(\psi(A)), \qquad A \in B(H).$$

By Sazonov's theorem, the characteristic functional $\tilde{\mu}_H$ is continuous with respect to a bilinear form $(Th,h)_H$ where T is an S-operator on H. Then we have

$$\begin{aligned}\tilde{\mu}(\xi) &= \int_E e^{i\langle x,\xi\rangle} d\mu(x)\\ &= \int_H e^{i(h,\psi^*\xi)} d\mu_H(h)\\ &= \tilde{\mu}_H(\psi^*\xi), \qquad \xi \in E',\end{aligned}$$

where ψ^* is the conjugate operator of ψ. Consequently, $\tilde{\mu}(\xi)$ is continuous with respect to the bilinear form

$$\langle\psi T\psi^*\xi,\ \xi\rangle = (T\psi^*\xi,\ \psi^*\xi)_H.$$

Let Ξ be the collection of all bilinear forms

$$B(\xi\ ;\ H,\ \psi,\ T) = \langle \psi T \psi^* \xi,\ \xi \rangle, \qquad \xi \in E',$$

where ψ is a continuous injection of a Hilbert space H into E and T is an S-operator on H. Then we show that Ξ defines an S-topology on E'.

Assume that a non-negative definite functional $\chi(\xi)$ on E' is Ξ-continuous. Then for every natural number n there exists a bilinear form $B_n(\xi) = B(\xi\ ;\ H_n,\ \psi_n,\ T_n)$ in Ξ such that

$$(1) \qquad |1 - \chi(\xi)| < \frac{1}{n}, \quad \text{if} \quad B_n(\xi) < 1, \quad \xi \in E'.$$

Since T_n is an S-operator on H_n,

$$\tilde{\gamma}'_n\ (h) = \exp[-\tfrac{1}{2}(T_n h,\ h)_{H_n}], \quad h \in H_n,$$

is the characteristic functional of a Gaussian measure γ'_n on H_n. Let γ_n be a Gaussian measure on E defined by $\gamma_n = \gamma'_n \circ \psi^{-1}$. Then we have $\tilde{\gamma}_n(\xi) = \tilde{\gamma}'_n\ (\psi_n{}^*\xi)$ and $B_n(\xi)$ is the covariance of γ_n. Let $\{X_n\}$ be an independent E-valued random sequence with the distribution $\{\gamma_n\}$. Then $X = \sum_n 2^{-n}\sigma_n X_n$ converges almost surely, where $\sigma_n = E[\|X_n\|]^{-1}$, $n = 1,2,3, \ldots$. In fact we have

$$E[\|X\|] \leqslant \sum_n 2^{-n}\sigma_n\ E[\|X_n\|] < +\infty.$$

Let γ be the distribution of X. Then γ is a Gaussian measure on E with the covariance

$$(2) \qquad B(\xi) = \sum_n 2^{-2n}\sigma_n^2 B_n(\xi).$$

Since, by Lemma 1, E is of cotype 2, there exists a continuous injection ψ of a Hilbert space H into E and an S-operator T on H such that

$$B(\xi) = \langle \psi T \psi^* \xi, \xi \rangle$$

$$= (T\psi^*\xi, \psi^*\xi)_H, \qquad \xi \in E',$$

(S.A. Chobanjan and V.I. Tarieladze [2], Corollary 4-1-2°), that is, $B(\xi)$ belongs to Ξ.

On the other hand, it is obvious from (1) and (2) that $\chi(\xi)$ is continuous with respect to the 'single' bilinear form $B(\xi) = B(\xi : H, \psi, T)$. For every $\varepsilon > 0$ there exists $\delta > 0$ such that

$$|1 - \chi(\xi)| < \varepsilon, \quad \text{if} \quad B(\xi) < \delta, \quad \xi \in E',$$

and $B(\xi)$ does not depend on ε.

Define a functional χ_H on the subspace $\psi^*(E')$ of H by

$$\chi_H(\psi^*\xi) = \chi(\xi), \quad \xi \in E'.$$

If $\psi^*\xi = \psi^*\eta$, $\xi, \eta \in E'$, then we have $B(\xi-\eta) = 0$ so that $\chi_H(\psi^*\xi) = \chi_H(\psi^*\eta)$. Therefore χ_H is well-defined and evidently non-negative definite. Since ψ is injective, $\psi^*(E')$ is dense in H and, by the uniform continuity, χ_H is extended to a non-negative definite functional on H continuous with respect to $(Th,h)_H$. Consequently, by Sazonov's theorem, there exists a probability measure μ_H on H with the characteristic functional χ_H.

Let μ be a probability measure on E defined by $\mu = \mu_H \circ \psi^{-1}$. Then we have

$$\tilde{\mu}(\xi) = \int_E e^{i\langle x,\xi\rangle} d\mu(x)$$
$$= \int_H e^{i(h,\psi^*\xi)_H} d\mu_H(h)$$
$$= \chi_H(\psi^*\xi) = \chi(\xi), \qquad \xi \in E'.$$

Therefore χ is the characteristic functional of a probability measure μ on E.

It is not difficult to show that Ξ defines a vector topology on E' by the same argument as above.

Thus we have proved the theorem.

§3. PROBABILITY MEASURE WITHOUT HILBERTIAN SUPPORT.

In this section we will construct explicitly a probability measure without Hilbertian support on ℓ_p, $1 \leq p < 2$.

Let $\{Y_n\}$ be an independent random sequence with the same Cauchy distribution $[\pi(1 + x^2)]^{-1}dx$, $\{\alpha_n\}$ be a sequence of positive numbers such that

(3) $$\sum_n \alpha_n < +\infty,$$

(4) $$\sum_n \alpha_n |\log \alpha_n| < +\infty,$$

and define $X_n = \alpha_n Y_n$, $n=1,2,3,\ldots$. Then by the Kolmogorov's three series theorem, $\sum_n |X_n|^p$ converges almost surely for all p, $1 \leq p < 2$, so that $\{X_n\}$ induces a probability measure μ_p on $\ell_p (1 \leq p < 2)$.

Fix an arbitrary p in $[1,2)$ and assume that there exists a continuous injection ψ of a Hilbert space H into ℓ_p such that $\mu_p(\psi(H)) = 1$. Then the characteristic functional

$$\tilde{\mu}_p(\xi) = \exp[-\sum_n \alpha_n |\xi_n|], \quad \xi = \{\xi_n\} \in \ell_q,$$

$(\frac{1}{p} + \frac{1}{q} = 1)$, is continuous with respect to a non-negative symmetric bilinear form

$$B(\xi) = \langle \psi T \psi^* \xi, \xi \rangle$$
$$= \sum_{m,n} a_{mn} \xi_m \xi_n, \qquad \xi \in \ell_q,$$

where T is an S-operator on H and (a_{mn}) is the corresponding matrix representation. Consequently there exists a positive constant K such that

$$(\sum_n \alpha_n |\xi_n|)^2 \leqslant K \sum_{m,n} a_{mn} \xi_m \xi_n, \qquad \xi \in \ell_q.$$

Let $\{\varepsilon_n\}$ be the standard Bernoulli sequence. Then we have

$$(\sum_n \alpha_n |\xi_n|)^2 = (\sum_n \alpha_n |\varepsilon_n \xi_n|)^2$$
$$\leqslant K \sum_{m,n} a_{mn} \varepsilon_m \xi_m \varepsilon_n \xi_n, \qquad \xi \in \ell_q,$$

and, estimating the mathematical expectations,

$$(\sum_n \alpha_n |\xi_n|)^2 \leqslant K \sum_n a_{nn} \xi_n^2, \qquad \xi \in \ell_q.$$

Since all α_n's are positive, all a_{nn}'s are also positive and we have

(5)
$$\sup\{(\sum_n \alpha_n |\xi_n|)^2 \ ; \sum a_{nn} \xi_n^2 < 1\}$$
$$= \sum_n \frac{\alpha_n^2}{a_{nn}} < K < +\infty.$$

On the other hand it is not difficult to show that $B(\xi)$ is the covariance of a Gaussian measure on ℓ_p (see the proof of Theorem), so that we have

(6)
$$\sum_n a_{nn}^{\frac{p}{2}} < +\infty,$$

(N. Vakhania [12]).

But the conditions (3), (4), (5) and (6) do not always hold simultaneously. For example, let

$$\alpha_n = n^{-\frac{p+2}{2p}}, \qquad n = 1,2,3, \ldots,$$

then, although $\{\alpha_n\}$ satisfies (3) and (4), there is no matrix (a_{mn}) that satisfies (5) and (6) at the same time. In fact by Hölder's inequality we have

$$+\infty = \sum_n \frac{1}{n} = \sum_n \alpha_n^{\frac{2p}{2+p}}$$

$$\leq \left(\sum_n \frac{\alpha_n^2}{a_{nn}}\right)^{\frac{p}{p+2}} \left(\sum_n a_{nn}^{\frac{p}{2}}\right)^{\frac{2}{p+2}}.$$

This shows that μ_p has no Hilbertian support.

BIBLIOGRAPHY

[1] Baxendale, P. (1976) Gaussian measures on function spaces. *Amer. J. Math.* Vol. 98, pp891-952.

[2] Chobanjan, S.A. and Trieladze, V.I. (1977) Gaussian characterization of certain Banach spaces. *J. Mult. Anal.* Vol. 7, pp183-203.

[3] Diallo, B. (1977) Correction and supplement to the article "On the Hilbert subspace of full measure of the space ℓ_p." *Zapiski Nauchnix Seminaroff LOMI.* Vol. 72, pp213-214. (in Russian)

[4] Dudley, R.M. (1967) The sizes of compact subsets of Hilbert space and continuity of Gaussian processes. *J. Func. Anal.* Vol. 1, pp290-330.

[5] Mandrekar, V. (1978) Characterization of Banach space through validity of Bochner theorem. Vector space measures and applications I, *Lecture Note in Math.* Vol. 644, Springer

pp314-326.

[6] Maurey, B. (1972) Espaces de cotype p, 0 < p $\leqslant$ 2. *Seminaire Maurey-Schwartz 1972-1973, Exposé 7.*

[7] Mouchtari, D. (1976) Sur l'éxistence d'une topologie du type de Sazonov sur un espace de Banach. *Seminaire Maurey-Schwartz 1975-1976, Exposé 17.*

[8] Parthasarathy, K.R. (1967) Probability measures on metric spaces. *Academic Press, N.Y.*

[9] Sato, H. (1969) Gaussian measure on a Banach space and abstract Wiener measure. *Nagoya Math. J.* Vol. 36, pp65-81.

[10] Sazonov, V. (1958) A remark on characteristic functionals. *Th. of prob. its Appl.* Vol. 3, pp188-192 in English translation.

[11] Smolyanov, O.G. and Uglanov, A.V. (1973) Every Hilbert subspace of a Wiener space has measure zero. *Matematicheskie Zametki.* Vol. 14, pp772-774 in English translation.

[12] Vakhania, N. (1965) Sur les répartitions de probabilités dans les espaces de suite numeriques. *C.R.Acad.Sc.Paris.* Vol. 260, pp1560-1562.

Vol. 551: Algebraic K-Theory, Evanston 1976. Proceedings. Edited by M. R. Stein. XI, 409 pages. 1976.

Vol. 552: C. G. Gibson, K. Wirthmüller, A. A. du Plessis and E. J. N. Looijenga. Topological Stability of Smooth Mappings. V, 155 pages. 1976.

Vol. 553: M. Petrich, Categories of Algebraic Systems. Vector and Projective Spaces, Semigroups, Rings and Lattices. VIII, 217 pages. 1976.

Vol. 554: J. D. H. Smith, Mal'cev Varieties. VIII, 158 pages. 1976.

Vol. 555: M. Ishida, The Genus Fields of Algebraic Number Fields. VII, 116 pages. 1976.

Vol. 556: Approximation Theory. Bonn 1976. Proceedings. Edited by R. Schaback and K. Scherer. VII, 466 pages. 1976.

Vol. 557: W. Iberkleid and T. Petrie, Smooth S^1 Manifolds. III, 163 pages. 1976.

Vol. 558: B. Weisfeiler, On Construction and Identification of Graphs. XIV, 237 pages. 1976.

Vol. 559: J.-P. Caubet, Le Mouvement Brownien Relativiste. IX, 212 pages. 1976.

Vol. 560: Combinatorial Mathematics, IV, Proceedings 1975. Edited by L. R. A. Casse and W. D. Wallis. VII, 249 pages. 1976.

Vol. 561: Function Theoretic Methods for Partial Differential Equations. Darmstadt 1976. Proceedings. Edited by V. E. Meister, N. Weck and W. L. Wendland. XVIII, 520 pages. 1976.

Vol. 562: R. W. Goodman, Nilpotent Lie Groups: Structure and Applications to Analysis. X, 210 pages. 1976.

Vol. 563: Séminaire de Théorie du Potentiel. Paris, No. 2. Proceedings 1975–1976. Edited by F. Hirsch and G. Mokobodzki. VI, 292 pages. 1976.

Vol. 564: Ordinary and Partial Differential Equations, Dundee 1976. Proceedings. Edited by W. N. Everitt and B. D. Sleeman. XVIII, 551 pages. 1976.

Vol. 565: Turbulence and Navier Stokes Equations. Proceedings 1975. Edited by R. Temam. IX, 194 pages. 1976.

Vol. 566: Empirical Distributions and Processes. Oberwolfach 1976. Proceedings. Edited by P. Gaenssler and P. Révész. VII, 146 pages. 1976.

Vol. 567: Séminaire Bourbaki vol. 1975/76. Exposés 471–488. IV, 303 pages. 1977.

Vol. 568: R. E. Gaines and J. L. Mawhin, Coincidence Degree, and Nonlinear Differential Equations. V, 262 pages. 1977.

Vol. 569: Cohomologie Etale SGA 4½. Séminaire de Géométrie Algébrique du Bois-Marie. Edité par P. Deligne. V, 312 pages. 1977.

Vol. 570: Differential Geometrical Methods in Mathematical Physics, Bonn 1975. Proceedings. Edited by K. Bleuler and A. Reetz. VIII, 576 pages. 1977.

Vol. 571: Constructive Theory of Functions of Several Variables, Oberwolfach 1976. Proceedings. Edited by W. Schempp and K. Zeller. VI, 290 pages. 1977

Vol. 572: Sparse Matrix Techniques, Copenhagen 1976. Edited by V. A. Barker. V, 184 pages. 1977.

Vol. 573: Group Theory, Canberra 1975. Proceedings. Edited by R. A. Bryce, J. Cossey and M. F. Newman. VII, 146 pages. 1977.

Vol. 574: J. Moldestad, Computations in Higher Types. IV, 203 pages. 1977.

Vol. 575: K-Theory and Operator Algebras, Athens, Georgia 1975. Edited by B. B. Morrel and I. M. Singer. VI, 191 pages. 1977.

Vol. 576: V. S. Varadarajan, Harmonic Analysis on Real Reductive Groups. VI, 521 pages. 1977.

Vol. 577: J. P. May, E_∞ Ring Spaces and E_∞ Ring Spectra. IV, 268 pages. 1977.

Vol. 578: Séminaire Pierre Lelong (Analyse) Année 1975/76. Edité par P. Lelong. VI, 327 pages. 1977.

Vol. 579: Combinatoire et Représentation du Groupe Symétrique, Strasbourg 1976. Proceedings 1976. Edité par D. Foata. IV, 339 pages. 1977.

Vol. 580: C. Castaing and M. Valadier, Convex Analysis and Measurable Multifunctions. VIII, 278 pages. 1977.

Vol. 581: Séminaire de Probabilités XI, Université de Strasbourg. Proceedings 1975/1976. Edité par C. Dellacherie, P. A. Meyer et M. Weil. VI, 574 pages. 1977.

Vol. 582: J. M. G. Fell, Induced Representations and Banach *-Algebraic Bundles. IV, 349 pages. 1977.

Vol. 583: W. Hirsch, C. C. Pugh and M. Shub, Invariant Manifolds. IV, 149 pages. 1977.

Vol. 584: C. Brezinski, Accélération de la Convergence en Analyse Numérique. IV, 313 pages. 1977.

Vol. 585: T. A. Springer, Invariant Theory. VI, 112 pages. 1977.

Vol. 586: Séminaire d'Algèbre Paul Dubreil, Paris 1975–1976 (29ème Année). Edited by M. P. Malliavin. VI, 188 pages. 1977.

Vol. 587: Non-Commutative Harmonic Analysis. Proceedings 1976. Edited by J. Carmona and M. Vergne. IV, 240 pages. 1977.

Vol. 588: P. Molino, Théorie des G-Structures: Le Problème d'Equivalence. VI, 163 pages. 1977.

Vol. 589: Cohomologie l-adique et Fonctions L. Séminaire de Géométrie Algébrique du Bois-Marie 1965–66, SGA 5. Edité par L. Illusie. XII, 484 pages. 1977.

Vol. 590: H. Matsumoto, Analyse Harmonique dans les Systèmes de Tits Bornologiques de Type Affine. IV, 219 pages. 1977.

Vol. 591: G. A. Anderson, Surgery with Coefficients. VIII, 157 pages. 1977.

Vol. 592: D. Voigt, Induzierte Darstellungen in der Theorie der endlichen, algebraischen Gruppen. V, 413 Seiten. 1977.

Vol. 593: K. Barbey and H. König, Abstract Analytic Function Theory and Hardy Algebras. VIII, 260 pages. 1977.

Vol. 594: Singular Perturbations and Boundary Layer Theory, Lyon 1976. Edited by C. M. Brauner, B. Gay, and J. Mathieu. VIII, 539 pages. 1977.

Vol. 595: W. Hazod, Stetige Faltungshalbgruppen von Wahrscheinlichkeitsmaßen und erzeugende Distributionen. XIII, 157 Seiten. 1977.

Vol. 596: K. Deimling, Ordinary Differential Equations in Banach Spaces. VI, 137 pages. 1977.

Vol. 597: Geometry and Topology, Rio de Janeiro, July 1976. Proceedings. Edited by J. Palis and M. do Carmo. VI, 866 pages. 1977.

Vol. 598: J. Hoffmann-Jørgensen, T. M. Liggett et J. Neveu, Ecole d'Eté de Probabilités de Saint-Flour VI – 1976. Edité par P.-L. Hennequin. XII, 447 pages. 1977.

Vol. 599: Complex Analysis, Kentucky 1976. Proceedings. Edited by J. D. Buckholtz and T. J. Suffridge. X, 159 pages. 1977.

Vol. 600: W. Stoll, Value Distribution on Parabolic Spaces. VIII, 216 pages. 1977.

Vol. 601: Modular Functions of one Variable V, Bonn 1976. Proceedings. Edited by J.-P. Serre and D. B. Zagier. VI, 294 pages. 1977.

Vol. 602: J. P. Brezin, Harmonic Analysis on Compact Solvmanifolds. VIII, 179 pages. 1977.

Vol. 603: B. Moishezon, Complex Surfaces and Connected Sums of Complex Projective Planes. IV, 234 pages. 1977.

Vol. 604: Banach Spaces of Analytic Functions, Kent, Ohio 1976. Proceedings. Edited by J. Baker, C. Cleaver and Joseph Diestel. VI, 141 pages. 1977.

Vol. 605: Sario et al., Classification Theory of Riemannian Manifolds. XX, 498 pages. 1977.

Vol. 606: Mathematical Aspects of Finite Element Methods. Proceedings 1975. Edited by I. Galligani and E. Magenes. VI, 362 pages. 1977.

Vol. 607: M. Métivier, Reelle und Vektorwertige Quasimartingale und die Theorie der Stochastischen Integration. X, 310 Seiten. 1977.

Vol. 608: Bigard et al., Groupes et Anneaux Réticulés. XIV, 334 pages. 1977.

Vol. 609: General Topology and Its Relations to Modern Analysis and Algebra IV. Proceedings 1976. Edited by J. Novák. XVIII, 225 pages. 1977.

Vol. 610: G. Jensen, Higher Order Contact of Submanifolds of Homogeneous Spaces. XII, 154 pages. 1977.

Vol. 611: M. Makkai and G. E. Reyes, First Order Categorical Logic. VIII, 301 pages. 1977.

Vol. 612: E. M. Kleinberg, Infinitary Combinatorics and the Axiom of Determinateness. VIII, 150 pages. 1977.

Vol. 613: E. Behrends et al., L^P-Structure in Real Banach Spaces. X, 108 pages. 1977.

Vol. 614: H. Yanagihara, Theory of Hopf Algebras Attached to Group Schemes. VIII, 308 pages. 1977.

Vol. 615: Turbulence Seminar, Proceedings 1976/77. Edited by P. Bernard and T. Ratiu. VI, 155 pages. 1977.

Vol. 616: Abelian Group Theory, 2nd New Mexico State University Conference, 1976. Proceedings. Edited by D. Arnold, R. Hunter and E. Walker. X, 423 pages. 1977.

Vol. 617: K. J. Devlin, The Axiom of Constructibility: A Guide for the Mathematician. VIII, 96 pages. 1977.

Vol. 618: I. I. Hirschman, Jr. and D. E. Hughes, Extreme Eigen Values of Toeplitz Operators. VI, 145 pages. 1977.

Vol. 619: Set Theory and Hierarchy Theory V, Bierutowice 1976. Edited by A. Lachlan, M. Srebrny, and A. Zarach. VIII, 358 pages. 1977.

Vol. 620: H. Popp, Moduli Theory and Classification Theory of Algebraic Varieties. VIII, 189 pages. 1977.

Vol. 621: Kauffman et al., The Deficiency Index Problem. VI, 112 pages. 1977.

Vol. 622: Combinatorial Mathematics V, Melbourne 1976. Proceedings. Edited by C. Little. VIII, 213 pages. 1977.

Vol. 623: I. Erdelyi and R. Lange, Spectral Decompositions on Banach Spaces. VIII, 122 pages. 1977.

Vol. 624: Y. Guivarc'h et al., Marches Aléatoires sur les Groupes de Lie. VIII, 292 pages. 1977.

Vol. 625: J. P. Alexander et al., Odd Order Group Actions and Witt Classification of Innerproducts. IV, 202 pages. 1977.

Vol. 626: Number Theory Day, New York 1976. Proceedings. Edited by M. B. Nathanson. VI, 241 pages. 1977.

Vol. 627: Modular Functions of One Variable VI, Bonn 1976. Proceedings. Edited by J.-P. Serre and D. B. Zagier. VI, 339 pages. 1977.

Vol. 628: H. J. Baues, Obstruction Theory on the Homotopy Classification of Maps. XII, 387 pages. 1977.

Vol. 629: W.A. Coppel, Dichotomies in Stability Theory. VI, 98 pages. 1978.

Vol. 630: Numerical Analysis, Proceedings, Biennial Conference, Dundee 1977. Edited by G. A. Watson. XII, 199 pages. 1978.

Vol. 631: Numerical Treatment of Differential Equations. Proceedings 1976. Edited by R. Bulirsch, R. D. Grigorieff, and J. Schröder. X, 219 pages. 1978.

Vol. 632: J.-F. Boutot, Schéma de Picard Local. X, 165 pages. 1978.

Vol. 633: N. R. Coleff and M. E. Herrera, Les Courants Résiduels Associés à une Forme Méromorphe. X, 211 pages. 1978.

Vol. 634: H. Kurke et al., Die Approximationseigenschaft lokaler Ringe. IV, 204 Seiten. 1978.

Vol. 635: T. Y. Lam, Serre's Conjecture. XVI, 227 pages. 1978.

Vol. 636: Journées de Statistique des Processus Stochastiques, Grenoble 1977, Proceedings. Edité par Didier Dacunha-Castelle et Bernard Van Cutsem. VII, 202 pages. 1978.

Vol. 637: W. B. Jurkat, Meromorphe Differentialgleichungen. VII, 194 Seiten. 1978.

Vol. 638: P. Shanahan, The Atiyah-Singer Index Theorem, An Introduction. V, 224 pages. 1978.

Vol. 639: N. Adasch et al., Topological Vector Spaces. V, 125 pages. 1978.

Vol. 640: J. L. Dupont, Curvature and Characteristic Classes. X, 175 pages. 1978.

Vol. 641: Séminaire d'Algèbre Paul Dubreil, Proceedings Paris 1976–1977. Edité par M. P. Malliavin. IV, 367 pages. 1978.

Vol. 642: Theory and Applications of Graphs, Proceedings, Michigan 1976. Edited by Y. Alavi and D. R. Lick. XIV, 635 pages. 1978.

Vol. 643: M. Davis, Multiaxial Actions on Manifolds. VI, 141 pages. 1978.

Vol. 644: Vector Space Measures and Applications I, Proceedings 1977. Edited by R. M. Aron and S. Dineen. VIII, 451 pages. 1978.

Vol. 645: Vector Space Measures and Applications II, Proceedings 1977. Edited by R. M. Aron and S. Dineen. VIII, 218 pages. 1978.

Vol. 646: O. Tammi, Extremum Problems for Bounded Univalent Functions. VIII, 313 pages. 1978.

Vol. 647: L. J. Ratliff, Jr., Chain Conjectures in Ring Theory. VIII, 133 pages. 1978.

Vol. 648: Nonlinear Partial Differential Equations and Applications, Proceedings, Indiana 1976–1977. Edited by J. M. Chadam. VI, 206 pages. 1978.

Vol. 649: Séminaire de Probabilités XII, Proceedings, Strasbourg, 1976–1977. Edité par C. Dellacherie, P. A. Meyer et M. Weil. VIII, 805 pages. 1978.

Vol. 650: C*-Algebras and Applications to Physics. Proceedings 1977. Edited by H. Araki and R. V. Kadison. V, 192 pages. 1978.

Vol. 651: P. W. Michor, Functors and Categories of Banach Spaces. VI, 99 pages. 1978.

Vol. 652: Differential Topology, Foliations and Gelfand-Fuks-Cohomology, Proceedings 1976. Edited by P. A. Schweitzer. XIV, 252 pages. 1978.

Vol. 653: Locally Interacting Systems and Their Application in Biology. Proceedings, 1976. Edited by R. L. Dobrushin, V. I. Kryukov and A. L. Toom. XI, 202 pages. 1978.

Vol. 654: J. P. Buhler, Icosahedral Golois Representations. III, 143 pages. 1978.

Vol. 655: R. Baeza, Quadratic Forms Over Semilocal Rings. VI, 199 pages. 1978.

Vol. 656: Probability Theory on Vector Spaces. Proceedings, 1977. Edited by A. Weron. VIII, 274 pages. 1978.

Vol. 657: Geometric Applications of Homotopy Theory I, Proceedings 1977. Edited by M. G. Barratt and M. E. Mahowald. VIII, 459 pages. 1978.

Vol. 658: Geometric Applications of Homotopy Theory II, Proceedings 1977. Edited by M. G. Barratt and M. E. Mahowald. VIII, 487 pages. 1978.

Vol. 659: Bruckner, Differentiation of Real Functions. X, 247 pages. 1978.

Vol. 660: Equations aux Dérivée Partielles. Proceedings, 1977. Edité par Pham The Lai. VI, 216 pages. 1978.

Vol. 661: P. T. Johnstone, R. Paré, R. D. Rosebrugh, D. Schumacher, R. J. Wood, and G. C. Wraith, Indexed Categories and Their Applications. VII, 260 pages. 1978.

Vol. 662: Akin, The Metric Theory of Banach Manifolds. XIX, 306 pages. 1978.

Vol. 663: J. F. Berglund, H. D. Junghenn, P. Milnes, Compact Right Topological Semigroups and Generalizations of Almost Periodicity. X, 243 pages. 1978.

Vol. 664: Algebraic and Geometric Topology, Proceedings, 1977. Edited by K. C. Millett. XI, 240 pages. 1978.

Vol. 665: Journées d'Analyse Non Linéaire. Proceedings, 1977. Edité par P. Bénilan et J. Robert. VIII, 256 pages. 1978.

Vol. 666: B. Beauzamy, Espaces d'Interpolation Réels: Topologie et Géometrie. X, 104 pages. 1978.

Vol. 667: J. Gilewicz, Approximants de Padé. XIV, 511 pages. 1978.

Vol. 668: The Structure of Attractors in Dynamical Systems. Proceedings, 1977. Edited by J. C. Martin, N. G. Markley and W. Perrizo. VI, 264 pages. 1978.

Vol. 669: Higher Set Theory. Proceedings, 1977. Edited by G. H. Müller and D. S. Scott. XII, 476 pages. 1978.